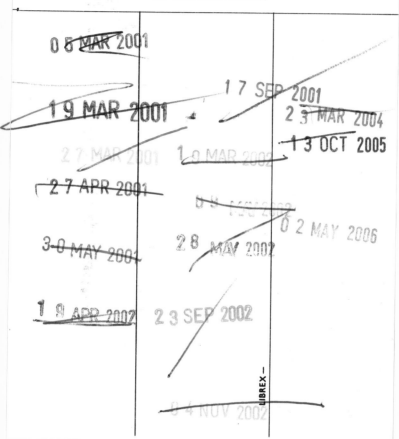

LOCUS OF CONTROL
Current Trends in Theory and Research

LOCUS OF CONTROL
Current Trends
in Theory and Research

HERBERT M. LEFCOURT
University of Waterloo

 LAWRENCE ERLBAUM ASSOCIATES, PUBLISHERS
1976   Hillsdale, New Jersey

DISTRIBUTED BY THE HALSTED PRESS DIVISION OF

JOHN WILEY & SONS

New York   Toronto   London   Sydney

Lawrence Erlbaum Associates, Inc., Publishers
62 Maria Drive
Hillsdale, New Jersey 07642

Distributed solely by Halsted Press Division
John Wiley & Sons, Inc., New York

Library of Congress Cataloging in Publication Data
Lefcourt, Herbert M
    Locus of control.

    Bibliography: p.
    Includes index.
    1. Control (Psychology)   I. Title.
BF632.5.L43   153   76-4895
ISBN 0-470-15044-0

Printed in the United States of America

Contents

Preface

Since the locus of control construct was introduced into the psychological literature in the early 1960s, there has been an overwhelming abundance of research pertaining to the perception of control. The task of researchers has thus become more difficult if overstimulated. One is rarely certain of being up to date on developments in related but slightly different areas near one's own core interests.

The purpose of this book is to help bring some order to this impending chaos. Each area reviewed represents a major focus of interest that has preoccupied different groups of investigators. The scope of each review, however, is limited to the degree that only those studies deemed central or seminal by this author are given detailed attention. Other investigations not considered to be as salient or stimulating are not discussed in detail, if at all. Consequently, this book is not totally comprehensive except insofar as it presents overviews of most of the primary thrusts that have been evident in locus of control research.

Given the organization of this book around topics that have been central to researchers, it may be used as a reference text by investigators or as an introduction to given areas and problems for potential researchers seeking to become familiar with the state of knowledge of locus of control research. Though there have been several volumes in which related topics such as hopelessness, helplessness, personal causation, and causal attribution have been given extended coverage, this book is unique in focusing largely upon the locus of control construct as it developed within social learning theory. Though several review articles have been published, this volume is the first to give extended treatment to this important construct within a social learning framework.

In this book readers will find a detailed description of research investigations and discussions of the social and personal ramifications of the findings reported. It will become evident to the reader that psychological research can be addressed to larger philosophical questions concerning the nature of man, and this may prove to be of interest for graduate and advanced undergraduate students in psychology who express cynicism about the relevance of psychology for understanding the human condition.

Most generally this book will be of interest to psychologists and other behavioral and social scientists who are concerned with personality and psychopathology. The approach taken in this book derives directly from the excellent working example of the Boulder-model orientation in clinical psychology that was presented by the Ohio State University Clinical Psychology faculty of the late 1950s. The congruence between research and clinical activity, which is at the core of the Boulder model, was convincingly illustrated among that faculty. Consequently, the writer can express appreciation to Julian B. Rotter and the late George A. Kelly, Alvin Scodel, and Shephard Liverant. In addition, I note my appreciation of the late Vaughn Crandall who first introduced me to social learning research during my undergraduate years.

Finally, this writer would like to offer appreciation to each of several individuals: to my wife Barbara who helped me to develop an "internal critic" for evaluating my own writing, to Elizabeth Hogg and Carol Sordoni, my research assistants, who performed an assortment of roles that facilitated the process of book writing, and to the wonderful University of Waterloo secretaries who somehow or other were able to familiarize themselves with some woeful handwriting.

HERBERT M. LEFCOURT

LOCUS OF CONTROL
Current Trends in Theory and Research

1
Locus of Control:
Understanding the Concept

INTRODUCTION

Man has reached a stage in his history which surrounds him with challenges to his very survival. Faced with problems of overpopulation and the consequent overcrowding, discomfort, and environmental pollution, the individual can easily come to feel overburdened, irritable, and hopeless. The title of the Broadway musical "Stop The World I Want To Get Off" is a common distillation of the thoughts of many an urban dweller. One of the most disturbing accompaniments of population density has been the sharp increase in incidents of violence with the resulting insecurity to which urban dwellers have become accustomed. Robberies, assaults, sometimes with no apparent motive, murders, and rape have become commonplace in sections of many large cities. Citizens are purchasing weapons for protection of their homes, and certain communities have established protective wire or electronic screening with guarded entrances to prevent the possibility of encountering marauders. In New York City the number of large guard dogs is said to be enjoying such an increase that a street sanitation problem has developed.

If it were not so serious, much of the world today could be construed as a bad joke. Tempted by the repeated evidence of man's cruelty to man throughout his history it is possible to question whether man is likely to survive in the relatively free state within which he lives today, crowded closer together with others, and dependent upon some belief in the ephemeral good will or conscience within himself and others for his daily feelings of safety.

Writers such as Arthur Koestler (1967) and Hannah Arendt (1963) and psychological researchers such as Stanley Milgram (1963) lead us to believe that man is indeed a creature strange enough to be capable of destroying himself and

1

his planet for what would seem to be trivial reasons: a desire for social approval or loyalty to one's in-group, for example.

Concurrent with these negative views of man's nature some writers have stressed the need for man to surrender his myths of freedom and sovereignty. Norman Cousins, the editor of *Saturday Review,* has repeatedly called upon nations to acknowledge the fruitlessness of national sovereignty and to submit to world law and order. Sovereignty is said to be an anachronism in a time of accelerating speed of communication, travel, and weapons development, and the consequent increasing rate of crisis occurrence.

On the more personal level, the behaviorist B. F. Skinner argues for the need of man to surrender his myths of freedon and will (Skinner, 1971). Despite the oppressive tone of Skinner's book, *Beyond Freedom and Dignity,* it has apparently become a best seller, something unique for the writings of psychologists. Whereas Norman Cousins' formulations of world order seem appealing to persons of liberal sensibilities, Skinner's derogation of free will does not. Carl Rogers, Rollo May, Arthur Koestler, and others view these Skinnerian pronouncements as totalitarian in nature (*Time,* 1971).

In Skinner's thinking, man must relinquish his belief in freedom and self-determination and come to accept the fact that he is controlled by forces outside of himself. With such acceptance Skinner believes that man will become more responsive to those controlling forces which reinforce what is more naturally acceptable to humans. Today's relatively random world in which normlessness and unpredictability prevail would thus cease to be as man would avail himself of rewards for more orderly and mannerly behavior. The chance elements of childhood and social experiences that can come to produce psychotic assassins and deviates of all manner would be eliminated and most men would become altruistic and pleasant to one another. Would that such a world could be!

The thesis of this book is that uncertainties and variations in personal experience sometimes *produce* deviates but also *produce* ingenious and creative individuals. Unique and innovative minds grow among those who can come to perceive differences between others and themselves, and who continue to hold the assumption that they are free agents, the makers of their own fates.

This book focuses upon research that has been conducted in psychological laboratories and in field settings concerning the effects of an individual's perception of control. Whether people, or other species for that matter, believe that they are actors and can determine their own fates within limits will be seen to be of critical importance to the way in which they cope with stress and engage in challenges. In other words, what Skinner believes to be an irrelevant illusion will be shown to be a very relevant illusion—one that seems to be central to man's ability to survive, and, what is more, to enjoy life.

The position presented in this book is not offered with the enthusiasm engendered by confident rebuttal and refutation. Skinner's wish for a more orderly and mannerly world is shared by many if not most individuals. It is paradoxical, however, that the very surrender of the *belief* in free will advocated by Skinner as a step in the direction of a less violent world can be viewed as a source of increased violence, especially that of a prosocial kind. While the data and arguments presented in this book are generally counter to Skinner's position regarding the myth of individual freedom, they are presented with an awareness that the maintenance of the belief in individual freedom is not without considerable cost.

The work presented in these pages argues that man must come to be more effective and able to perceive himself as the determiner of his fate if he is to live comfortably with himself. It is through the very abnegation of perceived self-direction and the surrender to indomitable forces that man comes to join with others like himself to commit horrendous acts upon dissimilar others. An ironic twist, however, inheres in the likelihood that as we encourage individualism and privacy, we generate more loneliness, discontent, and personal misery among those less advantaged and beget more antisocial criminality. Ecologists have made us aware that whichever way man turns he creates disorder, if not of one sort then of another.

EMPIRICAL DEMONSTRATIONS OF THE CONCEPT

A series of psychological experiments that have immediate relevance to urban malaise will serve to introduce the concept of perceived control. David Glass, Jerome E. Singer, and their colleagues (Glass & Singer, 1972; Glass, Reim, & Singer, 1971; Glass, Singer, & Friedman, 1969; Reim, Glass, & Singer, 1971) conducted a series of investigations concerned with the effects of noise upon tasks requiring persistence and attention to details. In several of these experiments, subjects had to complete a set of simple tasks: number comparisons, addition, and letter finding. In the number comparisons task, subjects were asked to indicate whether the multidigit numbers in each set of a series of pairs were the same or different. The addition task simply required subjects to add sets of one and two digit numbers. The letter finding task required that subjects find the letter A in five words of a column of 41 words. These tasks, simple enough in themselves, were administered under four conditions of noise distraction (Glass, Singer, & Friedman, 1969).

An aversive noise was created by combining the sounds of two people speaking Spanish, one person speaking Armenian, a mimeograph machine, a desk calculator, and a typewriter to produce a composite, nondistinguishable roar. No doubt the roar was a good replica of general urban mayhem. The authors, all New York residents, were well experienced with such stimuli. However, it was not simply the fact of noise and its impact upon persons working at simple tasks that was of interest. Rather, the concern was with the effects of the predictability and controllability of the noise. In this study each of four groups received a different combination of stimuli. One group was subjected to the noise at 110 decibels (loud noise) for 9 seconds at the end of every minute of the session. Another group received the loud noise but at random intervals and for random lengths of time. Two other groups received fixed and random noise but at a softer volume (56 dB).

As might have been expected the initial bursts of noise were effective in distracting most subjects. However, as the session progressed, subjects adapted to the noise, improving in performance on the simple tasks, and exhibiting lesser responsivity as assessed by physiological measurements. After this session, subjects were engaged in two further tasks during which there was no noise or like interference. One task was designed to evoke frustration. The second task, while routine, required caution and attentiveness.

The frustration task consisted of design copying in which subjects had to trace over all lines in each of four designs with certain restrictions: subjects were not to lift the pencil from the paper at any time and were not to trace over any line twice. Several copies of each design were available so that a subject could make as many attempts to succeed as he desired. Two of the diagrams were insoluble. Therefore, no matter how many times the subject attempted to succeed, he failed.

The dependent variable measured in this task was the number of times that subjects attempted to solve the insoluble designs. Subjects who had received noise at fixed intervals did not differ from a control group which had not been subjected to noise at all. Those subjects who had suffered the noise on a random schedule made considerably fewer attempts, and this was most pronounced when the random noise had been of the higher intensity.

The second task was a proofreading job. Ten errors were included on each of seven pages of an essay. Misspellings, grammatical errors, and so on were to be located during a fifteen minute period. Performance on the proofreading task was measured in terms of the percentage of omitted errors. While the results were somewhat less significant than for the tracing task, subjects who had experienced random noise made more omissions than those who had received fixed interval noise. Loud random noise was associated with the highest percentage of omitted errors, whereas the softer fixed interval noise was associated with the best proofreading performance.

While the intensity of noise had some effect on the subjects' performance, it paled in comparison to the effects of predictability. The implications of these results regarding predictability deserve close attention. If noise, or any aversive stimulus for that matter, were unanticipated, the shock value of the stimulus would no doubt be augmented. Who has not found himself startled by soft but unexplainable sounds occurring in the night? The state of alertness and arousal thus engendered in the individual would probably be continuous if one were in a strange place where the meaning of noises was uncertain. In other words, if we do not know the significance of a noise it will become arousing; the perceiver will feel a need to become alert and ready—for anything, the limits of anticipation and apprehensiveness depending upon that person's imagination.

It is now appropriate to discuss the relationship between the meaning of a noise and the predictability of that noise. If we "know" a sound, such as the starting of a furnace motor, we know from where it originates and what consequences may be expected from that sound. The sequence is predictable; nothing untoward is anticipated. The sound will not change in pitch or intensity. If unusual changes do occur, we would become suspicious of the working of that furnace and summon a repairman. In short, predictability is a major facet of knowing something. An overly predictable person generates little arousal or interest and often creates boredom in others. The consistency and reliability of noise in the investigation described above instructs the subject that subsequent changes in volume and timing of the noise are unlikely; there need be little apprehension of a sudden increased intensity. As regularity is perceived, the subject can also ready himself, slowing down in his work efforts when he anticipates the onset of noise. He can, therefore, avoid interruptions by not letting himself be caught unawares and distracted in the midst of an activity.

Implicit in this discussion of predictability is the element of control. If we know the ordinary sounds of the furnace, then we know what must be done to it and when. If another person is predictable, then we have a good idea of how one must act with him to cause certain effects. In each instance, predictability allows us some sense of confidence that we can act to create desirable effects. In being forced to hear predictable noise we may stop work and wait until it ceases, or steel ourselves for the onset, minimizing our own responses to the noise. We are not as helpless as we might otherwise be since we can do something to minimize the impact of predictable noise. It is this perception of the ability "to do something" that gives rise to the concept of perceived control.

In a second investigation reported by Glass *et al.* (1969), the effect of perceived control was more directly examined. All of the subjects received loud randomly occurring noise; the most aversive and debilitating condition found in the first study. The major difference between the first and second studies was that, in the second investigation, half of the subjects were provided with a button that would enable them to terminate the noise. These subjects were instructed as to the use of the button, but were encouraged to use it only if the noise became too much for them to bear. In essence, the subjects were provided with a modern analog of control—the off switch.

Subjects with access to the off switch tried almost five times the number of insoluble puzzles and made significantly fewer omissions in proofreading than did their counterparts who were given no such option for controlling the aversive stimulation. These differences were obtained despite the fact that none of the subjects who had potential control actually used it. The mere knowledge that one can exert control, then, serves to mitigate the debilitating effects of aversive stimuli.

From the investigations reported by Glass *et al.* (1969), it may be concluded that when an aversive event is predictable its effects are minimized. This may result from the opportunity that regularity creates for us either to schedule our efforts so as to avoid interruption, or to prepare our sensory apparatus so as to be less sensitive to the disturbing event. Second, control of the termination of the aversive stimulus diminishes the impact of that stimulus, perhaps, by eliminating the fear that "things can get worse" and even beyond endurance. Conceivably it is *fear of unendurable pain* that is debilitating to the cognitive processes of individuals undergoing the experience of unpredictable and uncontrollable but only mild to moderate levels of pain and irritation.

Extrapolation from these conjectures regarding fear of pain to the suffering of anxiety may provide an explanation of well-known clinical phenomena. One may ask whether an individual would experience anxiety if he had ways of avoiding events that offered threat to him. Perhaps, "la belle indifference," the beautiful indifference of hysterical women toward their functional disabilities during the Victorian era, derived from a sense of inner content. Conversion reactions, through which neurologically impossible paralyses occur, often created "secondary gains" for patients in that they would be allowed to avoid "aversive experiences" such as sexual intercourse. It is tempting to suggest that, for many hysterical patients who frequented psychoanalysts' offices during the first decades of this century, the hysterical symptoms were a "button," a device by which it became possible to terminate threatening events. Similarly, therapy as it is practiced today often seems directed toward helping a patient find a better "button," one that enables him to more comfortably approach and cope with threats.

Such conjecture would seem premature if it were not for the reliability and consistency of research on perceived control. Glass, Reim, and Singer (1971) were able to replicate the "access to a button" phenomenon described above when subjects themselves were not in direct control, but could ask a partner to press the button and terminate the noise for them. With shock as the aversive stimulus, Staub, Tursky, and Schwartz (1971) found that subjects who were allowed to administer shock to themselves and to select the level of intensity of that shock reported less discomfort at higher levels of shock and endured stronger shocks than did paired subjects to whom shocks were administered passively. However, when all subjects were given a second series of shock trials administered without subject control, the group that previously experienced control declined in their tolerance for the shocks, rated lower intensity levels as being more uncomfortable, and endured less shock than previously. On the other hand, no changes were found among subjects who had not experienced control.

The Glass and Singer experiments exhibit the effects of control upon task performance under aversive conditions. The investigation by Staub *et al.* (1971) reveals similar shifts in self-reports in that the reported aversive quality of a stimulus decreased when subjects exercised control over that stimulus. These results are congruent with findings reported by Pervin (1963) that subjects preferred predictable, self-controlled shock administration to unpredictable, experimenter-controlled conditions. Parallel findings with regard to acknowledged anxiety in each condition was obtained. Other investigators (Corah & Boffa, 1970; Haggard, 1943) have found that stress, as measured by physiological changes, was reduced when subjects were able to control the onset and termination of aversive stimulation.

It seems evident, then, that the exercise of control and the ability to predict the occurrence of aversive stimuli have an ameliorating effect upon the recipient. Pain-producing stimuli prove less painful and disruptive to individuals who can predict and control those stimuli. Such findings have been obtained for different types of response measures, that is, performance, self-report, and physiological indices. To further document the significance and pervasiveness of the effects of perceived control, research findings among other species are presented in the subsequent section.

The Response to Aversive Stimulation among Nonhumans

A classic study concerned with the ramifications of control was reported by Mowrer and Viek (1948). In this seminal and theoretically prophetic experiment, Mowrer and Viek were able to show that rats exhibited less fear of an aversive stimulus when they could exercise control in terminating it. For each of 15 days, 20 food-deprived rats were offered a bit of food after they had been placed in an experimental cage. Ten seconds after the food was taken by the rat a shock was delivered through a grid at the floor of the cage.

For half of the sample the shock was left on until the rat lept into the air. If the rat did not eat within 10 seconds after food was presented, this failure to eat was regarded as an inhibition. The food was subsequently withdrawn and the shock applied 10 seconds later. One group of 10 animals could terminate the shock through leaping; a second group of 10 were passively yoked to the first group; that is, each member of what was referred to as the "shock-uncontrollable" group was paired with one rat from the shock-controllable group and received shock for whatever length of time his controlling pair received it on each respective day. The crucial dependent variable in this study was the number of inhibitions recorded in each group for each of the 15 successive days.

The 10 rats in the shock-controlling group had an overall total of only 16 inhibitions [mean (M) = 1.6] throughout the experiment. In contrast, the shock-noncontrolling or helpless animals produced 85 inhibitions (M = 8.5), a difference that was strongly significant. In addition, the helpless group notably increased in the incidence of inhibition from the first three days (0, 1, 3) to the last three days (8, 8, 8) whereas the shock-controlling group did not inhibit at all until the third day at which time only one inhibition was noted. Only on four days did more than one inhibition occur for the shock-controlling group, and there were but two inhibitions on three of these days, and three were obtained on the other day. The helpless group, on the other hand, produced five or more inhibitions on 12 of the 15 days.

These results support the hypothesized effect of control rather well. When the rats could terminate the shock, they exhibited less fear-related behavior. Since all of the animals were hungry, the shock-controlling rats can be said to have been acting more in their own interests in continuing to eat the proffered food. The inhibition of eating in the helpless rats, on the other hand, can be construed as a maladaptive response. Frozen with fear, these animals were unable to eat despite their hunger. One might say that the helpless rats "lost their will" or the active response of defending their self-interests. Though such terms often inspire a facetious response from psychologists, constructs such as hope, helplessness, and will have been recently resurrected in a number of works to help account for the persistence of human activity despite sometimes overwhelming adversity (May, 1969; Menninger, 1963).

Mowrer (1950) seemed keenly aware of some of the far-reaching ramifications of his work: "Perhaps we have isolated here, in prototype, one of the central reasons why human beings so universally prize freedom and why threats to freedom, under a totalitarian regime, are anxiety-producing [p. 472]." Mowrer also extrapolated from his results to the rather mundane but human experience of concern with illness:

One is ill and suffering from pain and inconvenience. The physician arrives, diagnoses the difficulty, prescribes treatment, and intimates that in a day or two one will be quite hale again. It is unlikely that the examination or the ensuing exchange of words has altered the physical condition of the patient in the least; yet he is likely to "feel a lot better" as a result of the doctor's call. What obviously happens in such instances is that initially the patient's physical suffering is complicated by concern lest his suffering continue indefinitely or perhaps grow worse. After a reassuring diagnosis, this concern abates; and if, subsequently, the same ailment recurs, one can predict that it will arouse less apprehension than it did originally [Mowrer, 1950, pp. 472–473].

Not to be caught anthropomorphizing, Mowrer attempted to explain the shock-controlling rat's relief as deriving from anticipatory motoric movements associated with leaping, the shock-terminating response. Later investigators examining the effect of control among animals have been less hesitant to employ concepts such as hopelessness and futility in discussing the behavior of animals.

Among the more dramatic investigations concerned with the loss of control among infrahumans is a study reported by Curt Richter (1959). Richter observed some unanticipated sudden deaths among his laboratory rats as they underwent different procedures. For example, after whisker cutting, an occasional animal would display a strange corkscrewing motion that eventually ended in death. Richter did not initially draw any inferences from the strange occurrences until these unexplained deaths became more frequent in later experimentation.

In a study concerned with swimming endurance at varying water temperatures, Richter found that a few animals whose whiskers had been trimmed would swim around excitedly for a few seconds in a turbulent bath, dive to the bottom apparently in search of escape, and then, after swimming around for a short time below the surface, would suddenly stop and die. Most fascinating were the autopsy results which revealed that the animals had not drowned. Richter had found rats capable of swimming for up to 81 consecutive hours under ideal conditions. Consequently, the short-lived attempt at swimming and sudden death presented a mysterious challenge for a curious investigator.

Up to this point, Richter had found whisker trimming to be a major determinant of this sudden death phenomenon. While whisker trimming produced sudden death among wild rats (street and farm bred), it seemed less deleterious to the tame laboratory-bred variety. Additionally, several wild rats whose whiskers were not trimmed died suddenly after being placed in the swimming jars so that whisker trimming was obviously not the sine qua non determinant of this strange occurrence. Ultimately, Richter concluded that handling per se seemed to be the primary cause of death among wild rats. Handling, while producing arousal among wild rats, also prevented any instrumental activity that could result in escape from the aversive experience. But if the rats were allowed to escape just once, the sudden death phenomenon was eliminated. Richter (1959) observed:

> Interesting evidence showing that the phenomenon of sudden death may depend on emotional reactions to restraint or confinement in glass jars comes from the observation that after elimination of hopelessness the rats do not die. On several occasions we have immersed rats in water and promptly removed them. The animals quickly learned that the situation was not actually hopeless and so became aggressive and tried to free themselves or escape and showed no signs of giving up. Such conditioned rats swam on the average 40 to 60 hours or more. Once freed from restraint in the hand or confinement in the glass jar, speed of recovery is remarkable. A rat that would certainly have died in another minute or two becomes normally active and aggressive in only a few minutes [p. 309].

Richter concluded that neither restraint alone nor whisker trimming kills a rat. Rather, death results from a combination of responses to various stresses occurring in rapid succession, which generate a sense of hopelessness in the animal. Richter expressed this in the following manner:

> This sudden-death phenomenon may however be considered also as a reaction at a much higher level of integration. The situation of these rats is not one that can be resolved by either fight or flight—it is rather one of hopelessness: being restrained in the hand or in the swimming jar with no chance of escape is a situation against which the rat has no defense. Actually, such a reaction of apparent hopelessness is shown by some wild rats very soon after being grasped in the hand and prevented from moving. They seem literally to give up [pp. 308–309].

In other words, the wild animal whose repertoire of behavior consists of rapid movements such as biting, running, and jumping is suddenly bereft of any adequate and ready response for coping with stressful demands. And yet, if rescued once, the animal, as it were, seems to learn that if it only perseveres, all is not hopeless—torture is not infinite. Richter's rats did not learn an instrumental response. Rescue was a fortuitous event, and the animal could not have learned what it could do to prevent a recurrence of his good fortune.

Some learning theorists, such as Guthrie, might suggest that removal from the swimming jar would be a reward for the last response made at the time of rescue. Expectancy theorists, on the other hand, might contend that the animal enjoyed a revival of hope through learning that there was an end and limit to the aversive stimulation. Consequently, if the rat were able to persist in swimming for rather lengthy periods of time, then survival was possible. Since swimming itself could not terminate their duress, the animals may simply be said to have learned to endure while waiting for the end of confinement.

Hopelessness in Psychiatric Settings

Endurance stimulated by hope of relief is familiar in medical settings. Patients must often wait as wounds heal, sometimes with little or no certainty that therapeutic healing will actually occur. While waiting for recovery, patients amuse themselves and keep actively engaged enough so that the desire to regain health is maintained. Deaths have been ascribed to a "lessened will to live" or to giving up in the face of improbable recovery.

The writer was witness to one such case of death due to a loss of will within a psychiatric hospital. A female patient, who had remained in a mute state for nearly ten years, was shifted along with her floor mates to a different floor of her building while her unit was being redecorated. The psychiatric unit where the patient in question had been living was known among the patients as the "chronic hopeless" floor. In contrast, the first floor to which the patient was moved was most commonly occupied by patients who held privileges, including the freedom to come and go on the hospital grounds and the surrounding streets. In short, the first floor was an exit ward from which patients could anticipate discharge fairly rapidly.

Patients temporarily moved from the third floor were given medical examinations prior to the move, and the patient in question was judged to be in excellent medical health though still mute and withdrawn. Shortly after moving to the first floor, the patient surprised the ward staff by becoming socially responsive and, within a two week period, she ceased being mute and was actually becoming gregarious. As fate would have it, the redecoration of the third floor unit was soon completed, and all previous residents were returned to it. Within a week after she had been returned to the "hopeless" unit, the patient, like the legendary Snow White who had been aroused from a living torpor, collapsed and died. The subsequent autopsy revealed no pathology of note, and it was whimsically suggested at the time that the patient had died of despair.

Cases such as this could be cast aside as selected and unrepresentative anecdotes if they were not so commonplace. Richter cited a variety of determinants of sudden, unexplained death which have included voodoo, hexes, fright, the sight of blood, and hypodermic injections. Unaccountable deaths have been reported among persons in good health who have attempted to commit suicide though they had barely scratched the surface of their skin or had only ingested a few aspirin tablets. Richter (1959) concluded: "Some of these instances seem best described in terms of hopelessness—literally a giving up when all avenues of escape appeared to be closed and the future holds no hope [p. 311]."

Kobler and Stotland (1964) have reported the outbreak of a series of suicides within a psychiatric hospital in which staff conflicts had resulted in the patients losing hope that they could ever improve. While this example could be more reasonably interpreted as an instrumental response for terminating a hopeless situation, the "passive" surrender of chronically ill patients may likewise represent a determined act. The patient gives in to what he sees as his inevitable decline and becomes apathetic, which then results in lessened responsiveness to his felt needs. The positive contribution of self-participation and involvement in medical healing has been examined with some success by Rue Cromwell and his colleagues (1968).

We will return to a discussion of some of these investigations in later chapters. At present, we shall draw our attention to two other groups of investigations which have implicated the importance of control among other nonhuman primates.

Escape Behavior in Dogs

Seligman, Maier, and Solomon (1969) have summarized the findings from a series of studies in which they investigated the effects of inescapable shock upon subsequent escape behavior of dogs. In most of these experiments (Overmier & Seligman, 1967; Seligman, 1968; Seligman & Maier, 1967; Seligman, Maier, & Geer, 1968), inescapable shock was administered by placing the dog in a cloth hammocklike harness so that the dog's legs hung below his body through four holes. A shock source was applied to the dog through electrodes that were taped to the footpads of the dog's hind feet. In this harness, the dogs were given a series of shocks, varying in duration and frequency in each experiment.

A second experimental procedure placed dogs in a two-way shuttle box in which escape or avoidance responding could be observed. In this unit the animal was exposed to a conditioned stimulus (often light dimming) after which an unconditioned stimulus (an electric shock) was administered through the grid floor. Whenever the animal crossed the shoulder high barrier in the center of the box, the shock was terminated. In contrast to the harness procedure in which an average stimulus was administered and the animal could do nothing that would alter the situation, the shuttle-box situation offered immediate control in that the shock could be eliminated altogether if the animal responded to the conditioned stimulus or warning signal by crossing the barrier.

Overmier and Seligman (1967) compared three groups of dogs that had received inescapable shock of varying frequency and duration with a group receiving no such treatment before entering the shuttle-box unit. Comparisons were made in latency of escape responses, the number of failures to escape shock, and the overall percentage of subjects that never escaped shock. The authors summarized their data as follows:

> Not only do statistical differences appear between the performances of Ss exposed to inescapable shock and those of unshocked Ss during subsequent instrumental avoidance training 24 hours later, but large qualitative differences also appear. The phenomenon we have reported is dramatic to observe. Whenever a S, which was not treated with inescapable shock, first received shock in the course of instrumental training, it typically barked, yelped, ran and jumped until it escaped. An S previously exposed to inescapable shock *initially* reacted to the first shock during instrumental training in much the same way. In contrast, however, it soon typically stopped vocalizing and moving in an agitated fashion and would remain silent until shock terminated. On succeeding trials, S would typically continue in a maladaptive pattern of behavior—not necessarily the same on each trial—and passively "accept" the severe pulsating shock [Overmier & Seligman, 1967, p. 30].

After experimenting with other relevant conditions, the authors posited that the source of interference in learning to escape from the shock in the shuttle box derived from helplessness learned in experiencing the inescapable shock. Subsequent experiments replicated the finding that inescapable shock interfered with learning instrumental behaviors for eliminating shock in the shuttle box and offered some clues as to how learned helplessness may be overcome. Prior experience with escapable shock immunizes dogs against the negative effects of later inescapable shock (Seligmen & Maier, 1967); that is, dogs that received a sequence of escapable and inescapable shock trials behaved very similarly in the shuttle-box unit as did animals that had only received escapable shock. In contrast, both of these groups differed greatly from those animals that received only inescapable shock. Seligman and Maier (1967) described their results as follows:

> Subjects which have had prior experience with escapable shock in the shuttlebox show more energetic behavior in response to inescapable shock in the harness. This contrasts with the interference effect produced by inescapable shock in Ss which have had no prior experience with shock or in Ss which have had prior experience with inescapable shock. Thus, if an animal first learns that its responding produces shock termination and then faces a situation in which reinforcement is independent of its responding, it is more persistent in its attempts to escape shock than is a naive animal [pp. 7—8].

The passivity engendering effect of inescapable shock was apparently eliminated in animals who previously had some successful experience of controlling shock. In a later quasi-therapy experiment, Seligman, Maier, and Geer (1968) attempted to overcome the interference effect of inescapable shock by encouraging successful behavior in the shuttle box. The barrier between sections of the shuttlebox was removed and the experimenter called out "here boy" to the dog through an open window on the side of the box opposite the side upon which the dog sat. If the dog responded to the call and crossed the lowered barrier shock, would have terminated and the dog would therefore have "fortuitously" learned of the escape route. One of the four subjects did respond to this treatment and began to escape shock in the shuttle box. The other three, however, had to be subjected to more strenuous approaches. With leashes attached to the dogs' collars, the animals were pulled across to the "safe side" on each trial of shock. The rationale was, of course, to force the subject to be exposed to the response-reinforcement contingency. The dog that responded to the simpler calling procedure began to escape reliably after 20 such trials. Dogs that had to be pulled across the barrier required 20, 35, and 50 such trials before escape became a reliable phenomenon. In short, the resistance to learning the simple contingencies of the shuttlebox was marked.

Seligman *et al.* (1968) concluded that the inappropriate passive acceptance of aversive stimuli among their canine subjects could be construed in terms of perceived lack of control over reinforcements.

The Development of Ulcers

Among the investigations discussed thus far, the phenomenon of perceived control of aversive stimuli would seem to be a central determinant of the manner in which one responds to these stimuli. Among humans, rats, and dogs, to have some instrumental response at one's disposal and to be able to perceive contingency between one's actions and the termination of an aversive stimulus seems to be a good and necessary thing. It serves us well, then, to refer to one investigation in which perceived effective control proved to be disadvantageous for the subject, whereas the passive recipient of aversive stimuli seemed to survive experimental treatment more adequately.

Brady, Porter, Conrad, and Mason (1958) found that monkeys pressing levers at a fast rate to avoid shocks developed ulcers. For this research, Brady devised a procedure whereby monkeys were yoked in an apparatus. One subject, euphemistically called the "executive" monkey, was trained to press a lever which would terminate a shock delivered to the animal's feet. Another monkey was connected electrically in series with the executive so that any shock received by one would also be received by the other. The executive could exercise control whereas the latter was a passive recipient. In this way both animals of a pair were subjected to the same shocks but one had "the button" and the other didn't.

After a number of days in this procedure, death occurred to each of four executives whose autopsies revealed extensive gastrointestinal lesions with ulceration. Sacrificed partners did not reveal any indications of such gastrointestinal complications. The findings in this study would encourage us to be a bit more temperate in judging the value of perceived control. However, in an elaborate replication of the "executive monkey" study, Weiss (1971) found evidence that ulcers were both more common, and the ulcerous lesions more extensive, among animals that have been deprived of control.

Although Weiss used a different species (rats) than Brady and his colleagues, the major reason for the reversal in findings was attributed to the method of selection of "executive" and "yoked" subjects by Brady *et al.* (1958). Subjects were *not* randomly selected for these roles. Rather, each pair of monkeys was administered a two to four hour avoidance pretest, and the monkey responding at the higher rate was always chosen as the executive.

The Weiss investigation was an improvement on the Brady study in several ways. First, Brady *et al.* (1958) employed a subject sample of only four pairs of monkeys where Weiss used 180 male rats. In addition, Weiss varied the manner in which the rats were to be cued to the onset of shock. From his overall findings, Weiss (1971) concluded:

> The present experiment showed that regardless of whether electric shock was preceded by a warning signal, by a series of warning signals forming, so to speak, an external clock, or by no signal at all, rats that could perform coping responses to postpone, avoid, or escape shock developed less severe gastric ulceration than matched subjects which received the same shocks but could not affect shock by their behavior. . . . the present results, in combination with earlier experiments, serve to establish that the beneficial effect of coping behavior in stressful situations is of considerable generality [p. 8].

CONCLUSIONS

On the basis of the evidence reviewed in this chapter on the response to aversive stimulation, perceived control seems to make a great difference. Pain or anxiety are not simply responses to aversive stimuli impinging upon our senses. Reactions to aversive stimuli are evidently shaped and molded by our perceptions of these stimuli and by perception of our ability to cope with these stimuli. While these conclusions are far from unique, it is remarkable that the findings appear similar across different species with different aversive stimuli and response measures. Behaviorists have often attempted to reduce differences between species through invoking universal principles such as reinforcement, but it is also possible to conclude that there are remarkable similarities among diverse species without reducing complex cognitive–perceptual systems to simple reinforcement assimilators. The perception of control would seem to be a significant determinant of the response to aversive events, regardless of species.

2
The Perception of Control
as an Enduring Attitude

INTRODUCTION

In the previous chapter, most of the investigations that were reviewed dealt with the effects of perceived control within circumscribed tasks or situations. For the most part, one would not anticipate that such effects would rapidly generalize to a variety of other tasks, nor that the organism under study would form some general self-conception on the basis of that limited experience. Nevertheless, the intractability of the dogs in Seligman's research and the surrender to death among Richter's animals do suggest that, for many organisms, the loss of perceived control is not taken as a simple isolated incident, but rather as a pervasive and profound reaction holding implications for the creature's judgment of his potential for survival.

When one acquaints himself with the massive literary outpourings regarding the lives of impoverished peoples, displaced persons, and of members of denigrated minority groups, a common characterization is that of abject helplessness and a sense of despair. To people who live in continuously adverse circumstances, life does not appear to be subject to control through their own efforts. Only through some outside intervention do events seem to be alterable, and such intervention is a rare occurrence. A most poignant example of fatalism is recorded by Oscar Lewis (1961). Manuel, one of the Sanchez family about whom Lewis writes, ruminates about the failure of his brief venture as a shoe manufacturer:

> To me, one's destiny is controlled by a mysterious hand that moves all things. Only for the select, do things turn out as planned; to those of us who are born to be tamale eaters, heaven sends only tamales. We plan and plan and some little thing happens to wash it all away. Like once, I decided to try to save and I said to Paula, "Old girl, put away this money so that some day we'll have a little pile." When we had ninety pesos

laid away, pum! my father got sick and I had to give all to him for doctors and medicines. It was the only time I had helped him and the only time I had tried to save! I said to Paula, "There you are! Why should we save if someone gets sick and we have to spend it all!" Sometimes I even think that saving brings on illness! That's why I firmly believe that some of us are born to be poor and remain that way no matter how hard we struggle and pull this way and that. God gives us just enough to go on vegetating, no? [Lewis, 1961, p. 171]

Manuel's fatalistic view is prototypic of the outlook of deprived individuals. In fact, it would seem to be foolish and intropunitive for many individuals to view themselves as active and self-controlling.

This position has been advanced by Gurin, Gurin, Lao, and Beattie (1969) who found that Negroes who blamed "the system" for their difficulties were more likely to exhibit innovative behavior than their more self-blaming counterparts. These authors contend that to belong to an economically deprived and socially denigrated minority means that one must inevitably experience difficulties and assaults upon his sense of dignity, and that it is to be expected that a person will not therefore blame himself for his fate. The important element to be considered for the present purpose, though, is the manner in which individuals come to terms with their difficulties and, more specifically, how they explain their difficulties to themselves.

In a compelling description of his odyssey through the Southeastern United States disguised as a Negro, John Howard Griffin (1962) explained the manner in which Negroes naturally learn to attribute causality to external rather than internal factors:

The Negro's only salvation from complete despair lies in his belief, the old belief of his forefathers, that these things are not directed against him personally, but against his race, his pigmentation. His mother or aunt or teacher long ago carefully prepared him, explaining that he as an individual can live in dignity, even though he as a Negro cannot. "They don't do it to you because you're Johnny—they don't even know you. They do it against your Negro-ness" [p. 48].

From the foregoing discussion one might infer that the development of external attributions for one's failures and difficulties is both veridical and functionally useful. The belief that one operates "the buttons" when in fact many of the more important environmental reinforcements are beyond one's own capability of control could seem delusional, and would, no doubt, incur some personal anguish. As indicated previously, however, to feel helpless, as if one were not an actor but merely a pawn, is likewise costly in terms of psychological well-being. Kurt Lewin (1940) wrote of the costs and benefits of externality in discussing the problems encountered by Jews as they emerged from a ghetto existence:

Using a term of dynamic psychology, we can say that the individual insofar as his Jewishness is concerned, becomes to a higher degree a separated whole than he was in the time of the Ghetto. At that time he felt the pressure to be essentially applied to the

Jewish group as a whole. Now as a result of the disintegration of the group he is much more exposed to pressure as an individual. The weakening of the pressure against Jews as a group since the Ghetto period has been accompanied by a development which has shifted the point of application of external forces from the group to the individual. . . . This shows the extent the previous behavior was due to the previous situation, a situation in which the individual was uncertain whether a disparagement of his work was attributable to its lack of merit or to the fact that its creator was a Jew. Even though the occasions for this uncertainty might have been rare, they could have the lasting effect of depriving the person of standards by which to measure the extent and limits of his ability, and in this way make him unsure of his own worth [p. 83].

In other words, when an individual is deprived of his sense of self-determination he is less able to learn about himself from his own experiences; he is less able to develop a definite measure of his own worth. It should come as no surprise, therefore, to discover that Negroes, whose socialization encourages a fatalistic outlook on life, have often been found to hold aberrant self-assessments. E. Franklin Frazier (1962), in his study of the Negro middle-class, wrote of the tendency among middle-class Negroes to develop a world of fantasy in which delusions of wealth, power, and prestige in white society were of paramount importance.

An Anecdotal Description of the Stability of Perceived Control

One of the more engaging descriptions of the inner life of the denigrated and depersonalized individual was contained in a letter to the Editor of *Harper's Magazine* (Scott, 1962). The letter was a rebuttal to an editorial by John Fischer, then editor, who had written of his despair at the rising rate of visible violence and apathy in Negro slum neighborhoods. Mr. Fischer had contended that blacks should organize into citizens' groups that would promote responsibility and self-control among themselves. The rebuttal, reprinted below, artfully presents the effects of pawn status upon an individual and emphasizes the inertia of such an adaptation:

The Negro and the Enlisted Man: An Analogy
Once in a while some of my friends who became commissioned officers in the service will sigh, shake their heads, and ask why I, who was born as proud as they, wasted three years as an Army GI. By now I've learned to shrug and feign puzzlement. For I know that soon enough—when the conversation turns to other topics and becomes sufficiently relaxed—they will reveal a total ignorance of certain experiences denied any middle-class American white except the peacetime Enlisted Man.

I was reminded of this in the July *Harper's* when Editor John Fischer wrote of the Negro's need to assume the responsibilities of first-class citizenship. About Mr. Fischer's service record I know nothing, but in that article he wrote as an officer. I too am a Northern liberal often discouraged by the refusal of Brother Tom to shape up. But as an ex-EM, I understand. In every dirty alley of Harlem, in every missed job of South Philly, I see myself lurking fatigue-clad behind Ft. Dix furnace rooms, relentless in my quest for the bonum minimum: the right to be bored in the way I choose.

Surely no other life open to the American white so closely resembles the Negro's—the world of Them and Us. They with their money, handsome uniforms, knowledge, organization, and (O God) their power. We with our anonymity ("D'ja ever notice how all EM look alike?"), dirt, encouraged stupidity, and uselessness. In such a world, only one weapon is available, but it is mighty: The Resolve to Live Up to Our Reputation. We are clods? dolts? animals? Very well. We shall be the cloddiest, most doltish animals on earth.

The discovery of this moral principle was for me the equivalent of Rousseau's glimpse of the Social Contract. Once I had found it, I could even divide all the EM I ever met into two camps—those who lived by The Resolve and those who rejected it. The former flourished; the latter (and maddeningly enough for me, a large number of these were fellow college boys) remained miserable whiners, petition passers, and demanders of their nonexistent rights—in short, second-rate citizens instead of first-rate slaves.

How did The Resolve operate? The details were so subtle and multitudinous as to defy codification. In fact, you had to be there. But if I had to say it in one sentence, this would do: Stay out of the stockade, but barely. Obey the letter, but defy the spirit. Rake leaves? All right, one leaf at a time. Sweep the floor, but sweep it with the somnambulant majesty of one who walks under water. Salute like a cheerful idiot. And, whether raking, sweeping, or saluting, give the appearance of doing your wretched, pathetic best.

Such incompetence will endear you to most officers, by confirming their prior notions of you as a comic stereotype; even better, it will make them feel necessary—an almost miraculous accomplishment. Only a few officers will penetrate the disguise, and from these you have nothing to fear. The truly wise will be amused and keep your secret. The merely clever will be maddened because they are helpless; the Army depends too heavily upon enlisted stupidity to risk punishing it.

I received dramatic proof of this very early—in my last week of basic training. Several thousand of us were slowly filing through a wooden shed, having our papers stamped and receiving our orders for permanent assignment. It was a gorgeous day. After all, perhaps the worst was over, and at the least we were going home for a week's leave. So with face shining and heart swelling, I passed through several stations without event.

It happened while I was standing in front of a warrant officer.

"Scott?" he said without looking up.

"Yes sir."

"Where are your glasses?"

Sudden, animal instinct told me something had gone wrong, but I took off my horn-rims and handed them to him.

"Negative. Where are your government issues?" He still hadn't looked at me.

"These are the only ones I have, sir."

Now he looked at me. "Where were you in the third week?"

I saw. In the third week I had been in the hospital with arthritic ankles; my boots had been a size too small. After returning to the company, I had noticed men wearing thick glasses with steel rims, almost like welders' goggles. Like a fool, I had sneered, thinking they had turned gung ho and bought PX glasses. Now with my stomach sliding into my shoes I could only repeat, "These are the only glasses I have, sir."

He wasn't looking at me any more. He was stamping my papers HOLD! "You'll be delayed a week. Your leave's cancelled. Get out of line. Next!"

I stumbled outside and saw my Field First leaning against a wall, looking bored. "Sergeant," I stammered, "what happens to someone held over a week?"

As always, at the prospect of inflicting pain, his boredom vanished. "Oh we got a lotta nice things. Strippin' rifles . . . sandin' tables . . . KP . . ."

All around, my comrades were jostling and swearing and hinting at how their leave-time would be spent. I made my choice.

Turning to a member of my platoon, I whispered, "Let me have those damn steel glasses for a minute." I grabbed them, ran back and put on the glasses.

My God! They were for a stigmatism. My myopia was as bad as ever, and now the floor was about fifteen feet below and slightly slanting up toward the right. But I could see well enough to know I was now standing in front of the officer.

"Jackson?" he said, not looking up.

"No sir," I said in my best dirt-farmerese. "I ain' Jackson. Scott. You got m' over thur'n the HOLD pile."

He looked up at me and remembered. "You didn't have government-issue glasses."

"Well, sir, you got m' all confused"—chuckle—"I didn' know you mean' *Army* glasses. I got muh Army glasses. I got'm on raht *here*."

I could see him glaring up at me from ten feet below. He was trying to decide whether I was a criminal or a moron. All he had to do, of course, was to ask me to read something. (Read! I couldn't even walk!)

Then, still deciding, he looked behind me and saw the endless stream of cattle with serial numbers to be stamped. He glared at me once more and decided I was insignificant—very likely a moron, but definitely insignificant. He grabbed a pen as though it were a scalpel, scratched out HOLD!, threw my papers at me, and looked away.

"Next!"

I was free. A free moron.

Do I hear protests from officers and civilians that this was civic irresponsibility? Do they believe that EM should prove themselves with honorable behavior and work well done? Then they ignore the implications of a social structure as rigid as a bear trap, where no truly compensatory rewards or punishments exist and where work is infinite in magnitude and infinitesimal in worth. You finish Job A only to have Job B created especially for you. And should you finish Job Z, no one will make you General, or Lieutenant, or even Corporal.

No. The irresponsibility is not civic, but private and psychological. The price you pay for the joyful security of pretending to be what you are not, is to become what you pretend to be.

As a pre-Army undergraduate, I assumed that campus Veterans were merely more sardonic than I toward the Ivory Tower—and more sardonic because they were more acquainted with Reality. Only later, myself a Vet in graduate school, did I know that our "cynicism" was the undirected nihilism of someone emerging from years of deception. We no longer knew ourselves. Our conditioned responses, once so useful and amusing in the Army, were now inappropriate. Our new situations demanded action, not endurance and sensual torpor; they demanded perceptions beyond the smug categories of four-letter expletives; they demanded, ultimately, that we remake ourselves in an almost forgotten image. Most of us did it, of course, suffering only the temporary pangs of rehabilitation. But we were left with an understanding that made it worthwhile.

Mr. Fischer is quite right. Every day the Negro's "irresponsibility" becomes more and more inappropriate. But before the Negro will see it, he needs an unequivocal, legal "discharge." And even then it will not be easy. The man who is *born* an EM cannot be rehabilitated. He needs metamorphosis. And unlike the fellow in Kafka, he can't do it overnight [Scott, 1962, pp. 16–21].*

SOCIAL OBSERVATIONS

The letter from Mr. Scott well illustrates how an individual may be reduced to a bumbling, if vigilant, serf, and describes approximately the accommodation to slavery and imprisonment. Stanley Elkins (1961) writes of the characteristics inherent in the American experience with slavery, and in the European experience with Nazi concentration camps, that result in the childlike, irresponsible behaviors of which Mr. Scott writes. Elkins defined the "Sambo" personality of Southern lore as an admixture of docility, irresponsibility, loyalty, laziness, humility, and dishonesty. The dominant plantation Negro was said to be "full of infantile silliness and his talk inflated with childish exaggeration [Elkins, 1961, p. 244]." Elkins also reviews the available literature regarding concentration camp life and concludes that, there too, childlike dependence was in evidence. In searching for the common ingredients that would create such infantilelike qualities, Elkins notes that total control of the slave or inmate lay directly in the hands of a master or guard. No alternative authorities existed and decisions regarding life or death resided in that one immediate authority:

> Everything—every vital concern—focussed on the SS: Food, warmth, security, freedom from pain, all depended on the omnipotent significant other, all had to be worked out within the closed system. Nowhere was there a shred of privacy; everything one did was subject to SS supervision. The pressure was never absent. It is thus no wonder that the prisoners should become "as children" [Elkins, 1961, p. 257].

In becoming "as children" one notable feature of Elkins' (1961) discussion centers on the loss of a sense of being a differentiated individual, a subject or actor: " . . . a more exquisite form of pressure lay in the fact that the prisoner had never a moment of solitude; he no longer had a private existence, it was no longer possible, in any imaginable sense, for him to be an individual [p. 250]."

The process through which the individual becomes depersonalized is described from first-hand observation by Bettelheim (1943):

> The author has no doubt that he was able to endure the transportation and all that followed, because right from the beginning he became convinced that these horrible and degrading experiences somehow did not happen to "him" as a subject, but only to "him" as an "object" [p. 431].

This split between subject and object as a device for maintaining oneself in the face of severe shock and strain is similar to the attributive processes mentioned previously by Griffin, Lewis, and Gurin and co-workers. In brief, when one is the recipient of shocking brutality, one experiences an unbelievable, that is, unpredictable and uncontrollable aversive event: the stuff of nightmares. Hence, one becomes, like the dreamer, at once the seer and the seen. The visible person is tortured while the active person retreats into a position of distant observer. Though the South, as portrayed by Griffin, or the ghetto by Lewin, is not equivalent in intensity to the total horror of early slavery or concentration camps, the parallels and similarities are obvious. Daily happenings were unpredictable and potentially dangerous, and the victims learned to be inconspicuous and overly responsive or acquiescent to external demands. Man the actor, the master of his fate, or simply the holder of the "button," absented himself and in his place was left the defensive pawn for whom good times were fortuitous and bad times inescapable.

EMPIRICAL FINDINGS

If these prosaic descriptions are as true as they would seem to be, then empirical data should substantiate the assumed link between deprivation, denigration, and perceived control. A contention of each of the writers discussed above is that helplessness, a perceived inability to effect one's fate meaningfully, is the natural response to deprivation and denigration and, in turn, is a source of immature and poor coping behavior. Among a number of investigations, one study conducted by the writer and a colleague (Lefcourt & Ladwig, 1965a) provides support for these contentions regarding black–white differences.

Lefcourt and Ladwig asked black and white reformatory inmates to respond to a questionnaire, referred to subsequently as the Internal–External Control of Reinforcement Scale (I–E Scale), that had been developed by Julian Rotter and his colleagues at the Ohio State University (Rotter, 1966; Rotter, Seeman, & Liverant, 1962; see Appendix). The scale consists of 23 forced-choice items along with six filler items designed to make the test purpose less obvious. Illustrative items are as follows:

I more strongly believe that:
  6. a. Without the right breaks one cannot be an effective leader.
     b. Capable people who fail to become leaders have not taken advantage of their opportunities.
  9. a. I have often found that what is going to happen will happen.
     b. Trusting to fate has never turned out as well for me as making a decision to take a definite course of action.
 23. a. Sometimes I can't understand how teachers arrive at the grades they give.
     b. There is a direct connection between how hard I study and the grades I get.

From a perusal of the test it is evident that the subject describes his own viewpoint by choosing between alternatives that reflect a fatalistic, external control viewpoint and those indicating a belief in his own ability to affect and to be in control of the events in his life. In the first comparison, the mean score for 60 black inmates was 8.97 [Standard Deviation (SD) = 2.97] and for 60 white inmates the mean was 7.87 (SD = 3.03). The resulting t test was significant (t =

2.00, $p < .05$), indicating that blacks acknowledged more external control than did whites. While the magnitude of the difference was not great, the fact that such a reliable difference could be obtained among a population that was homogeneous with regard to socioeconomic class and antisocial history indicates the meaningfulness of the obtained difference.

A second test of perceived control, Dean's Powerlessness Scale (Dean, 1969; see Appendix), was also administered to the same subjects. The Powerlessness measure consists of nine Likert scale items to which subjects indicate their degree of agreement or disagreement. Illustrative items are as follows:

> There is little or nothing I can do towards preventing a major "shooting" war.
> We're so regimented today that there's not much more room for choice even in personal matters.

On the Powerlessness scale, as with the I–E scale, a higher score indicates greater externality or fatalism. Again, the mean for black inmates ($M = 17.30$, $SD = 5.02$) was higher than the mean for white inmates ($M = 14.63$, $SD = 4.98$), and the difference was a reliable one ($t = 2.89$, $p < .01$). The extremity of the black's mean score on the Powerlessness measure is underscored by comparison with Dean's norms (Lefcourt & Ladwig, 1966). Based on a stratified sample of 384 males from Columbus, Ohio, the mean Powerlessness score was 13.65 ($SD = 6.10$). The black's mean was significantly higher than the normative mean ($t = 4.42$, $p < .001$). In contrast, the mean score for white inmates did not differ appreciably from the norm ($t = 1.20$, $p > .10$).

Another investigation by Esther Battle and Julian Rotter (1963) reported similar findings for black and white children with different measuring devices. Eighty sixth- and eighth-grade children selected on the basis of sex, social class, and race were tested on two measures of perceived control. One was the Bialer Locus of Control Questionnaire (Bialer, 1961; see Appendix), a 23-item "yes" or "no" questionnaire in which the subject attributes causality to himself or others. Examples of Bialer's items are as follows:

> Do you really believe a kid can be whatever he wants to be?
> When people are mean to you, could it be because you did something to make them be mean?
> When nice things happen to you, is it only luck?

A second measure was a "projective" device, a six-item cartoon test in which the child states what he would say in a series of lifelike situations involving attribution of responsibility. The items pictured in the six cartoons portrayed interactions to which the child had to respond. In the cartoons, another figure addressed the "subject" as follows:

> (1) How come you didn't get what you wanted for Christmas?
> (2) Why is she always hurting herself?

(3) When you grow up do you think you could be anything you wanted?

(4) Whenever you're involved something goes wrong!

(5) That's the third game we've lost this year.

(6) Why does her mother always "holler" at her? [Battle & Rotter, 1963, p. 485]

Performance on the cartoon test indicated that lower-class black children produced more responses coded as external ($M = 18.3, SD = 3.4$) than did either middle-class blacks ($M = 15.8, SD = 3.5$), or middle-class ($M = 15.0, SD = 4.4$) or lower-class white pupils ($M = 16.4, SD = 3.5$). The differences were significant between lower-class blacks and all of the other groups, while no differences were obtained among the other groups.

The Bialer scale was found to be correlated with the cartoon test ($r = -.47, p < .01$) which supported the belief that the cartoon test was providing a measure of perceived control. In addition, lower-class children expressed more externality on the Bialer scale than did middle-class children.

In general, then, the double derogation of class and caste was related to the expression of external control beliefs. These findings have been replicated with other populations, measuring devices, and procedures. Strodtbeck (1958), for example, has found that Jewish middle- and upper-class subjects express more of a sense of mastery than lower-class Italians. Most of the variance on Strodbeck's mastery scale was attributed to social class. For another example, Franklin (1963) reported a significant relationship between higher socioeconomic class and internality on the basis of a stratified national sample. With specific regard to race, blacks have been found to be more external or fatalistic in orientation than have whites with but few exceptions. Lessing (1969), Owens (1969), Shaw and Uhl (1969), Strickland (1972), and Zytkoskee, Strickland, and Watson (1971) have found such differences using ad hoc perceived control scales and published scales pertaining to control such as the Bialer, the I–E, and the Nowicki–Strickland (1973) scales (see Appendix).

On the other hand, a few studies have been reported in which class and race groupings were not reflected in measures of perceived control. Gore and Rotter (1963) failed to obtain social class differences on the I–E scale in a group of Southern Negro college students. Likewise, Katz (1967), and Solomon, Houlihan, and Parelius (1969) found no race differences in response to the Intellectual Achievement Responsibility scale (Crandall, Katkovsky, & Crandall, 1965; see Appendix), which measures perceived control in the academic settings of childhood. These investigators attributed their failure to obtain race related differences to the fact that classroom achievement situations may seem more amenable to control for blacks than the wider range of situations represented in other scales pertaining to control.

In one negative study in which the I–E scale was employed, Kiehlbauch (1968) found no differences between black and white reformatory inmates. This finding was obtained approximately six years after the investigation by Lefcourt and

Ladwig, who also utilized reformatory inmates. Differences between these studies may reflect the rise and fall of Martin Luther King, Malcolm X, the Kennedy brothers, and the tumultuous changes in race relations in the United States, and therefore may represent a real shift from fatalism to militancy, to the belief that one must act in one's own behalf or perish. Despite the occasional failure to obtain differences between racial groups on measures of locus of control and recent evidence to the effect that blacks may be shifting in their perception of control, it is notable that whenever differences are found, the black sample is more likely to espouse external fatalistic expectations. Likewise, members of the lower social economic class never exceed the more fortunate middle- and upper-class persons in statements of internality.

The most extensive investigation regarding group membership and the perception of control is, indubitably, a field study conducted by Jessor, Graves, Hanson, and Jessor (1968). These researchers explored in detail the relationship between economic and social factors contributing toward the development of deviant behavior. Rather than relying on a simple measure of socioeconomic status, the investigators explored the objective access to opportunity for three groups that comprised a community in the southwestern United States. Objective access was defined by the position of a subject on eight measures including age, marital status, language spoken in the home, occupation, education, religion, generation mobility, and social participation. Each position was conceptually linked to a location in the opportunity structure, an example of which is as follows:

> A young person has more objective access than an older one since he still has a chance to achieve presently unobtained goals in the future and since youth itself is responded to socially as an asset in socio-occupational life, whereas the older person's chances are diminishing, and age tends to be seen as a limitation in many areas of social life. Those elderly persons who have never married or who have lost their spouses have even less access to such life goals as successful child rearing, family participation, social involvement with others, and even occupational success [Jessor, Graves, Hanson, & Jessor, 1968, p. 126].

The total of the items, then, indicates the degree to which a person is in a position to secure valued ends. As had been predicted among the 221 community members surveyed, those with an "Anglo" identification had greater objective access to opportunity than did Spanish ($t = 10.63$, $p < .001$) and Indian ($t = 8.59$, $p < .001$) persons. When the three groups were compared with regard to scores on a modified form of Rotter's I–E scale, Anglos were found to be the most internal in contrast to the Spanish and Indians. The Spanish were more external than both Anglos ($t = 5.39$, $p < .001$) and Indians ($t = 3.78$, $p < .001$). Indians, on the other hand, did not differ significantly from Anglos ($t = 1.39$, n.s.). Of primary interest, however, is the fact that objective access and perceived control correlated strongly ($r = .50$, $p < .001$, $N = 221$). The more objective access to opportunity one had, the more potential control of one's fate did a person acknowledge. While ethnic group membership was also linked, the access to opportunity was more decidedly associated with perceived control.

CONCLUSIONS

In general, it may be concluded that perceived control is positively associated with access to opportunity. Those who are able, through position and group membership, to attain more readily the valued outcomes that allow a person to feel personal satisfaction are more likely to hold internal control expectancies. Blacks, Spanish-Americans, Indians, and other minority groups who do not enjoy as much access to opportunity as do the predominant caucasian groups in North American society are found to hold fatalistic, external control beliefs.

The anecdotes at the beginning of this chapter are supported by empirical data reported by various investigators using different measurement devices. Consequently, we may now speak of perceived control in more characterological terms than was the case in the first chapter. There, we addressed the question of whether having a sense of control would make a difference to an organism encountering an aversive event. In this chapter, we have found that deprived social position and severely punishing environments create a sense of fatalism along with infantile and regressive behavior. In the following chapter we will introduce a theoretical orientation which has helped to generate much of the interest in perceived control. In subsequent chapters, we use this theoretical orientation to examine the consequences of maintaining a sense of control as opposed to a fatalistic attitude.

3
Social Learning Theory: A Systematic Approach to the Study of Perceived Control

INTRODUCTION AND BACKGROUND

A quick perusal of the previous chapters makes it evident that the perception of control is not a provincial concern. Learning theorists with interests in the investigation of fear and stress, social psychologists who experiment with attribution processes, and clinical psychologists attempting to cope with their patients' helplessness and lack of confidence have all contributed to the growing literature dealing with the perception of control. The largest body of empirical data about perceived control, however, derives from Julian Rotter's social learning theory. It is in Rotter's theory that perceived control occupies a central place within a systematic formulation. While only a brief description of Rotter's views will be presented here, several books and articles contain detailed discussions of his theory (Rotter, 1954, 1955, 1960, 1971; Rotter, Chance, & Phares, 1972).

In Rotter's theory, a person's actions are predicted on the basis of his values, his expectations, and the situations in which he finds himself. The formulation for predicting behavior at a specific time and place is as follows:

$$BP_{x,S_1,R_a} = f(E_{x,R_a,S_1} \ \& \ RV_{a,S_1})$$

This formula reads: the potential for behavior x to occur in situation 1, in relation to reinforcement a, is a function of the expectancy of the occurrence of reinforcement a following behavior x, in situation 1, and the value of reinforcement a in situation 1.

This complicated formulation loses some of its obscure overtones when translated into a concrete example: the behavior of a young male collegian approaching an attractive female of the species and engaging in a series of appropriate courting behaviors, including eye contact, smiles, and mellifluous and witty conversation, is predictable with knowledge of the student's expectancy that his particular manner of approach will elicit reciprocal affective responses from this particular woman who he no doubt has been observing previously, at this place (cocktail party), and with the additional knowledge regarding the degree to which the woman's positive response is desired by him at this particular time.

In this formulation the importance of expectancies is not secondary to values. It is this equal emphasis upon value, expectancy of reinforcement, and situational specificity that makes Rotter's theory unique among learning theories which, more commonly, accentuate only the value or motive end of predictive formulas. For our present purpose of explicating the place of perceived control within social learning theory, we will shift our attention to Rotter's more general formula which reads as follows:

$$NP = f(FM \& NV)$$

The potentiality of occurrence of a set of behaviors that lead to the satisfaction of some need (need potential) is a function of both the expectancies that these behaviors will lead to these reinforcements (freedom of movement) and the strength or value of these reinforcements (need value). It is with the term freedom of movement that we approach the location of the locus of control construct in social learning theory.

Rotter defines freedom of movement as "the mean expectancy of obtaining positive satisfactions as a result of a set of related behaviors directed toward the accomplishment of a group of functionally related reinforcements. A person's freedom of movement is low if he has a high expectancy of failure or punishment as a result of the behaviors with which he tries to obtain the reinforcements that constitute a particular need [Rotter, 1954, p. 194]."

In essence, freedom of movement is a generalized expectancy of success resulting from man's ability to remember and reflect upon a lifetime of specific expectancy behavior–outcome sequences. Reverting back to our example of premating behavior, the male undergraduate, after a few years of comingling with the opposite sex, should establish a fairly stable estimate of success probability for eliciting feminine interest. This "stable estimate" would constitute his freedom of movement.

Perceived control is defined as a generalized expectancy for internal as opposed to external control of reinforcements. Like freedom of movement, it is an abstraction deriving from a series of specific expectancy behavior–outcome cycles. However, where freedom of movement concerns the likelihood of success, the generalized expectancy of internal versus external control of reinforcement involves a causal analysis of success and failure.

Let us return now to our student in his frozen state of approach toward an attractive female undergraduate. We may assume that he has accumulated a fund of information, deriving from similar past encounters, from which he has developed a generalized expectancy of success (freedom of movement). We may also assume that he understands why he arrives at the outcomes that he does. The student may have experienced many rebuffs in the past and developed a low freedom of movement regarding his ability to engineer satisfactory interactions with women. At the same time he, no doubt, acquired beliefs regarding the causes of his failures. He could attribute his failures to his own personal characteristics—homeliness, stupidity, or lack of assertiveness, wit, or conversational ease; or he may be in the habit of berating women for their perverse blindness in not recognizing his superlative qualities. "Women are all sluts," he may feel, "unable to appreciate the less than smooth, sensitive male."

Two such opposing interpretations of failure are examples of the internal and external poles of the locus of control dimension. Granted the generality of these expectancies for individuals holding such opposing views, we might conjecture that some individuals who attribute their failures to their own internal characteristics will become depressed following their failures, whereas others might busy themselves in improving their social manners in the hope of altering the outcomes of future encounters. The person who more readily attributed his social failures to the perversity of females, on the other hand, could perhaps become the perpetrator of sadistic acts upon women.

Habitual interpretations of failure may differ from one person to another, and success will not necessarily be interpreted in a similar way by different persons. Let us recreate our student, this time, in the guise of a friendly, outgoing person who has had little difficulty in approaching women for whatever purposes he has had in mind. He has a high freedom of movement for social encounters so that he engages in each new interaction with enthusiasm. While such an individual will, in all probability, describe himself as a socially successful person, there is no certainty that he will attribute his success to internal characteristics. Instead, he may actually feel puzzled about why women are attracted to him and blithely regard himself as being very lucky to be surrounded by such warm and generous women. Concepts such as "love at first sight" and the like may be part of his explanatory language.

These examples illustrate the fact that it is not the simple registering of success and failure experiences that is pertinent to the generalized expectancy of internal versus external control, but rather it is the interpretation of the cause of those experiences. Such an interpretation differs from the expectancy of success or failure in that it is concerned with our beliefs about how reinforcements are determined and should, therefore, provide an independent contribution along with freedom of movement and need value to the prediction of goal-directed activity.

In order to account more fully for our student's manner of approach to women we should like to know how much he values female company in contrast to other activities, how successful he believes he would be if he were to approach women in his characteristic ways, and whether he believes that the outcomes of his attempts are a function largely of his own behaviors, or are dependent on external circumstances such as "the nature of women." Thus, we can see that the internal processes which are often described by "self-oriented" theorists are implicit in social learning theory. With the locus of control construct, we are dealing with a person as he views himself in conjunction with the things that befall him and the meaning that he makes of those interactions between his self and his experiences.

In social learning terms the construct, perceived control, is referred to as a generalized expectancy of internal or external control of reinforcement. The formal terms, the generalized expectancy of internal control, refer to the perception of events, whether positive or negative, as being a consequence of one's own actions and thereby potentially under personal control. The generalized expectancy of external control, on the other hand, refers to the perception of positive or negative events as being unrelated to one's own behavior and thereby beyond personal control.

The source of interest in this construct did not begin simply with theoretical concerns but with problems encountered in psychotherapy. Rotter's social learning theory, unlike other similar approaches, has developed with a conjoint commitment to psychological research and to clinical practice. As Rotter (1966) describes it: " . . . the stimulus for studying such a variable has come from analysis of patients in psychotherapy . . . clinical analysis of patients suggested that while some patients appear to gain from new experiences or to change their behavior as a result of new experiences, others seem to discount new experiences by attributing them to chance or to others and not to their own behavior or characteristics [p. 2] ."

There is an apocryphal story about a patient engaged in psychotherapy which illustrates how an external control orientation can militate against improvement in therapy. A therapist trainee was said to be in despair about the progress of therapy with his patient. As a social learning-oriented clinician, the therapist had actively sought out significant persons in the patient's life who could deliver valued reinforcements to the patient. The passive, depressed, inadequate-feeling patient had often complained that no one cared about him or his achievements. The therapist subsequently instructed the patient's wife and employer, who were willing collaborators, in the "how and when" to reinforce the patient for his efforts. As a consequence, the wife and employer later reported that they were becoming rather pleased with the patient, and the patient himself reported that he had begun to feel more appreciated for his efforts. Nevertheless, he continued to appear to be depressed and suffering with feelings of inadequacy. When, in exasperation, the therapist confronted the patient with *the facts,* as he saw them, challenging the patient's perceptions, the patient countered with his tour de force—"True, things are improving—my wife and boss do attend to me more than they used to. They appreciate me more. *But, that's only because of you—you instructed them to be more appreciative of me."*

In other words, no matter the experiences one has, if they are not perceived as the results of one's own actions, they are not effective for altering the ways in which one sees things and consequently functions. Analogous therapy experiences have been reported by other writers with differing theoretical orientations and have been explained by such constructs as ego strength, self-esteem, inferiority feelings, hopelessness, "can-ness," and so on. From no other clinically relevant personality theory, however, has such an accumulation of research pertaining to perceived control developed.

In experiencing the kind of phenomenological data described above, the social learning theorist was almost bound to construct an expectancy variable to explain the patient's self-described inertia. The best theoretical statement introducing the expectancy of control construct was offered by Rotter (1966) in his review of research with the locus of control:

> In social learning theory, a reinforcement acts to strengthen an expectancy that a particular behavior or event will be followed by the reinforcement in the future. Once an expectancy for such a behavior-reinforcement sequence is built up the failure of the reinforcement to occur will reduce or extinguish the expectancy. As an infant develops and acquires more experience he differentiates events which are causally related to preceding events and those which are not. It follows as a general hypothesis that when the reinforcement is seen as not contingent upon the subject's own behavior that its occurrence will not increase an expectancy as much as when it is seen as contingent. Conversely, its nonoccurrence will not reduce any expectancy so much as when it is seen as contingent. It seems likely that, depending upon the individual's history of reinforcement, individuals would differ in the degree to which they attributed reinforcements to their own actions.
>
> Expectancies generalize from a specific situation to a series of situations which are perceived as related or similar. Consequently, a generalized expectancy for a class of related events has functional properties and makes up one of the important classes of variables in personality description. . . . A generalized attitude, belief, or expectancy regarding the nature of the causal relationship between one's own behavior and its consequences might affect a variety of behavioral choices in a broad band of life situations. Such generalized expectancies act to determine choice behavior along with the value of potential reinforcement. These generalized expectancies will result in characteristic differences in behavior in a situation culturally categorized as chance versus skill determined, and they may act to produce individual differences within a specific condition [Rotter, 1966, p. 2].

In this discussion Rotter explores the manner in which internal cognitive processes interact. The child is described as not assimilating new learnings if action–outcome sequences are perceived as being noncontingent; that is, he will not learn from his experiences unless he believes that these experiences are lawfully related to his own actions. If events are only randomly paired there would seem to be little reason for attending to them with an intent to learn. In fact, the term parataxic distortion was coined by Harry Stack Sullivan (1954) to connote just such an immature readiness to learn from what are only randomly connected events. It is evident, then, that the readiness to perceive contingency between one's actions and outcomes is an essential element in understanding how man comes to terms with his daily experience.

Some individuals, as was noted in Chapter 2, develop unshakeable beliefs that valued reinforcements occur only by chance, and that men are not the masters of their fates. In contrast, others may strongly believe that humans get their due desserts, that man is responsible for his fate. As suggested above, persons with such contrasting perspectives should differ considerably in the degree to which they are able to assimilate and learn from their experiences. The fatalists perceive no contingency between action and outcome, while those espousing internal control beliefs readily perceive such contingencies.

In later chapters, we will focus upon research on locus of control conducted within a social learning framework. In order to provide important background information, we will first examine the earliest social learning studies that helped to produce confidence in the belief that locus of control was indeed an important variable in behavior predictions.

EARLY EMPIRICAL INVESTIGATIONS

To test the assumption that the expectancy of control influences the way individuals behave, an obvious starting place is in the manipulation of specific task expectancies. In the first of several studies in which the expectancy of control was manipulated through task directions, Phares (1957) found evidence to support the contention that perceived control did matter. Phares developed two novel judgment tasks. In one, subjects were required to match sample paint patches of varying shades of gray, and in the other, they matched lines of slightly varying length. In both tasks, subjects had to judge whether a particular color or line was identical to a standard color or line that stood on a board some distance from them. To make these discriminations even more difficult, the standard color was presented against varying backgrounds and the standard lines were presented at different angles from true vertical. Consequently, the judging was so difficult that it was nearly impossible for subjects to know with any certainty whether they were correct, or whether the tasks were skill or chance determined. In light of such ambiguity, Phares instructed half of his subjects that the tasks were so difficult that success on them was more a matter of luck than of skill. The other half were informed that the tasks were a matter of skill and that some subjects were able to perform well at them.

Except for these different instructions, all subjects received the same series of experiences in the task. After each judgment, the subject was informed as to whether he was correct or not. This feedback was prearranged so that all subjects received the same sequence of "successes" and "failures." After each feedback, the subjects were asked to wager a number of poker chips to indicate their

expectancy of being correct on the succeeding trial. Of primary interest in research employing this "level of aspiration" paradigm is the manner in which subjects set their expectancies as a function of prior successes and failures. In the Phares experiment the interest centered on expectancy setting as a function of successes and failures when those reinforcements were perceived as controllable (skill) or random (chance).

As hypothesized, Phares found that reinforcement or feedback under "skill conditions" had a greater effect upon the manner in which subjects set their subsequent expectancies. Changes in the amount of chips wagered were significantly greater when subjects received skill than when they received chance instructions. Subjects receiving chance instructions more often "stood pat," which is a fair strategy in a gambling game. There was more shifting of expectancies among skill-instructed subjects, but chance-instructed subjects tended to make more "unusual shifts," that is, shifts that do not conform with one's previous experience. For example, chance-instructed subjects were more likely to increase their expectancies of success subsequent to failure and to decrease their expectancies of success following success. In Phares' experiment, more subjects who had received chance directions bet a lower number of chips after a "correct" and more chips after an "incorrect" judgment than did subjects in the skill condition. In other words, subjects given chance directions responded as if each success reduced a finite supply of luck, or as if, after failure, the likelihood of success increased.

Phares' investigation provided an early indication that knowledge of a subject's perception of control was useful for predicting the type of judgments he would make in response to success and failure in a given task. If a task is perceived as being solvable through the exercise of skill, then subjects will seriously use their experience with that task as a basis for making estimates as to their future likelihood of success or failure. On the other hand, if the results of performance are perceived as random occurrences, then subjects are more likely to ignore feedback and will begin behaving in a manner similar to that of a gambler.

A study by James and Rotter (1958) confirmed Phares' findings that perception of control predicted the manner in which people would respond to their performance outcomes. These investigators devised an experiment in which they were able to examine the effects of perceived control on the resistance to extinction after learning trials administered with partial versus 100% reinforcement. A time-honored research finding in psychology has been that persons who are partially reinforced in learning experiments (e.g., 50% reinforcement) will persist for a longer period of time after the occurrence of reinforcement has ceased than will persons who have received 100% reinforcement. Indeed, such findings have often been raised as an argument for the use of partial reinforcement in educational settings to generate prolonged efforts at learning without the teacher's close attention.

James and Rotter interpreted this partial reinforcement phenomenon in social learning terms, asserting that subjects who perceived the task as experimenter controlled, which is typically the case in learning experiments, would readily perceive a shift from 100% to 0% reinforcement as a change in the experiment. Subjects who had received partial reinforcement, on the other hand, would not discern the change in reinforcement rate as quickly, largely because the shift from partial to nonreinforcement was not as obvious.

In contrast, however, it was reasoned that if subjects believed that the reinforcement gained from their efforts was skill determined (internal), then different results would obtain. A shift from a 100% to a 0% reinforcement schedule might be interpreted more in terms of a sudden loss of the "touch," or some such internal attribute that could be compensated for with concerted effort. Such a phenomenon is familiar to sportsmen such as bowlers, in which "set" and fluidity of body movement are required, and without which skilled performance deteriorates. Largely because quick extinction would not be obtained after subjects had received 100% reinforcement during early trials in a skill-determined task, James and Rotter hypothesized that perceived control would prove to be a mediating variable for determining subjects' responses to partial versus 100% reinforcement schedules.

The subjects were presented with a card-guessing game in which they were to predict whether each subsequent card bore an X or an O. The cards were presented to each subject in a manually operated tachistoscope-like apparatus. From the subject's viewpoint each card dropped behind a closed shutter. After the subject stated his guess, the shutter opened and the card was revealed; the shutter then closed and another card dropped into place. From the experimenter's position there was a double-faced card with either an X or an O on it. When the subject stated his guess, the experimenter slid the card in the desired direction. In this way, the experimenter was able to control the outcome or reinforcement for each guess.

Before each trial, consisting of seven presentations and guesses, subjects registered their success prediction on a scale from 1 (failure) to 10 (success). The subjects were divided into four groups: 100%–skill, 100%–chance, 50%–skill, and 50%–chance. Skill and chance conditions were created through instructions, for example, "Before each trial I would also like you to estimate how well (skill) you feel you will do"; or, "how lucky (chance) you feel you will be." The subjects in the 100% reinforcement group were given 7 guesses on each of 10 trials. They were told that they were correct in 5 or 6 out of their 7 guesses on each trial. The 50% reinforcement group was given the same 10 trials, half of which were successes. After the tenth trial, extinction began and subjects were told that they were correct on 3, 2, or 1 out of the 7 guesses they had made within each trial. The instructions had previously defined such a low rate of correct guessing as due to chance; only trials in which 5 or more correct guesses

were made out of 7 were credited as better than chance. In addition subjects were penalized for inaccurate estimations and rewarded for accuracy in their total scores.

The results obtained by James and Rotter were consistent with their expectations. With chance instructions, the usual resistance to extinction differences were obtained between partial (29.05 trials) and 100% reinforcement conditions (15.55 trials), showing that partial reinforcement generated greater resistance to extinction (t = 4.43, p < .001). With skill instructions, however, the results were in the opposite direction, though failing to reach statistical significance (100%–skill, 22.90 trials; 50%–skill, 19.75 trials: t = 1.61, n.s.). The 100%–chance group extinguished more quickly (15.55 trials) than the 100%–skill group (22.90 trials), producing a t = 2.99, p<.01; and the 50%–skill group extinguished more quickly (19.75 trials) than the 50%–chance group (29.05 trials), t = 3.50, p < .001).

Thus, James and Rotter were able to demonstrate that expectancies regarding the controllability of outcomes make a considerable difference in the way subjects will construe changes in their experience. Extinction trials in a learning experiment may be perceived as failures by a subject which necessitate an increase in effort. On the other hand, the onset of extinction trials may be taken as evidence regarding the experimenter's behavior, signaling that the task has been changed. In the latter circumstance, subjects may even assume that it is their "awareness" that is under investigation so that it is the "right" or more intelligent choice to cease their efforts. These conflicting interpretations would seem to derive from the subject's perception of the task as one that does or does not test competence and, therefore, is or is not controllable through one's own efforts.

The greatest impetus for interest in locus of control derived from the creation of an assessment device, as is often the case in personality research. In Phares' (1957) investigation, a short Likert-type scale was used as a measure of locus of control. Phares' scale contained 13 items stated as external control-oriented attitudes and 13 as internal control-oriented attitudes. In Phares' study in which skill and chance sets were created by the use of instructions, he found that the 13 items stated as external control-oriented attitudes produced low-level correlations, bordering on statistical significance: individuals with external attitudes behaved in a fashion similar to subjects who had received chance instructions. That is, subjects who espoused external control attitudes made more unusual shifts of expectancies and yet made less and smaller shifts in expectancies that were congruent with their experiences than did subjects who had scored low on the 13 external control items.

Subsequently, James (1957) developed a larger scale, referred to as the James–Phares scale, which contained 26 items based on those which seemed to be the most useful from Phares' study. James also found low but significant correlations between his measure of locus of control and his subjects' responses to success and failure. Subjects characterized as holding an external control orientation exhibited smaller increments and decrements in expectancy statements following success and failure, and produced more unusual shifts than did internal control-oriented persons.

These early studies demonstrated that specific expectancies of control were manipulable experimentally and that generalized expectancies of internal–external control were assessable with paper and pencil devices. In subsequent chapters, it will become evident to the reader that these results opened up a research area that has rapidly mushroomed beyond its originator's most vivid expectations. Since the first introductory review of locus of control research (Rotter, Seeman, & Liverant, 1962), enough new and varied studies have appeared to stimulate several long review articles (Joe, 1971; Lefcourt, 1966a, 1972; Phares, 1973; Rotter, 1966) and these reviews were in need of updating almost before the print had dried. The remaining chapters are devoted to the attempt to catch up to this rapidly expanding research literature.

4

Locus of Control and the Resistance to Influence

INTRODUCTION AND BACKGROUND

In the wake of World War II Western man found himself in a state of dumb-foundedness. As the gates of concentration camps opened and the vestiges of a ruined humanity emerged, one fantasy of basic innocence and honor in human nature was shattered. People were led to ask how a "civilized" nation could descend into such bestiality and barbarism. Some of the aggrieved and self-righteous awaited revenge and judgment to occur at the Nuremburg trials. There, the victors expected confirmation of their beliefs, that each Nazi officer would be found to be corrupt, guilty, and personally evil. However, the trials failed to satisfy the prosecutions' desire for simple justice and elucidated some sinister implications for man. When asked, "How was it possible that all you honorable generals could continue to serve a murderer with such unquestioning loyalty?", one general replied that it was "not the task of a soldier to act as Judge over his supreme commander. Let history do that or God in heaven [Arendt, 1963, p. 101]."

Since those times, we have seen the question of responsibility for atrocities recur in many forms and in different circumstances. Hannah Arendt's *Eichmann in Jerusalem: A Report on the Banality of Evil* (1963) exposed to us at length a man who clearly portrayed the fact that the most horrendous acts derive more from obedience or compliance to social order than from sadistic impulse; thus the subtitle of Arendt's volume.

During his recollection of a meeting of SS and Nazi elite convened to deliberate about the "Final Solution," the mass murder of millions of Jews, Eichmann reflected upon the waning of his sense of repulsion at the thought of such violence. Seeing that other civil servants were enthusiastic about the espoused plans, he recalled that, "At that moment, I sensed a kind of Pontius

Pilate feeling, for I felt free of all guilt. Who was he to judge? Who was he to have his own thoughts in this matter? [Arendt, 1963, p. 101]":

> As Eichmann told it, the most potent factor in the soothing of his own conscience was the simple fact that he could see no one, no one at all, who actually was against the Final Solution. He did encounter one exception, however, which he mentioned several times, and which must have made a deep impression on him. This happened in Hungary when he was negotiating with Dr. Kastner over Himmler's offer to release one million Jews in exchange for ten thousand trucks. Kastner, apparently emboldened by the new turn of affairs, had asked Eichmann to stop "the death mills at Auschwitz" and Eichmann had answered that he would do it "with the greatest pleasure" but that, alas, it was outside his competence and outside the competence of his superiors—as indeed it was [Arendt, 1963, p. 103].*

In an attempt to describe the atmosphere of those times, Eichmann and several witnesses spoke of the legal web within which victims and captors alike were trapped:

> Eichmann . . . at least dimly realized that it was not an order but a law which had turned them all into criminals. The distinction between an order and the Führer's word was that the latter's validity was not limited in time and space, which is the outstanding characteristic of the former. This is also the true reason why the Führer's order for the Final Solution was followed by a huge shower of regulations and directives, all drafted by expert lawyers and legal advisers, not by mere administrators; this order, in contrast to ordinary orders, was treated as a law [Arendt, 1963, p. 133].*

In other words, the horrors perpetrated by the Nazi officials were legitimate, conforming to acceptable standards, and men like Eichmann felt that it was not for them to question, since what they were asked to do was legitimate. Evil in Nazi Germany had lost the aura by which most people recognize it, a unique quality of temptation to violate standards; and the common man, such as Eichmann, no longer felt able to judge what was right on some transcendent scale of values.

Confusion regarding legitimacy, moral judgment, and abnegation of a sense of moral integrity is no longer viewed as an aberrant phenomenon. With today's rapid shifting of perspectives it is difficult for individuals to retain an abiding sense of morality. All too many examples of man's capricious nature abound.

Stanley Milgram (1963, 1965) conducted a series of experiments in the early 1960s which indicate what history makes obvious, that harmful behavior can occur on demand and that such a happenstance is not uniquely characteristic of but a few odd social groups. Milgram, in the name of science, required subjects to administer a series of increasingly severe electric shocks to the hand of a respectable looking middle-aged man. While no actual shocks were delivered, to the naive subject it appeared as if he was administering painful shocks to the victim.

---

*From *Eichmann in Jerusalem* by Hannah Arendt. Copyright © 1963 by Hannah Arendt. Reprinted by permission of The Viking Press, Inc.

Milgram varied several elements in the experimental situation in the hope of deterring subjects from a too easy compliance. While some variability in subjects' behavior was obtained, the overall results were such as to cause Milgram (1965) to conclude:

> With numbing regularity, good people were seen to knuckle under the demands of authority and perform actions that were callous and severe. Men who are in everyday life responsible and decent were reduced by the trappings of authority, by the control of their perceptions, and by the uncritical acceptance of the experimenter's definition of the situation into performing harsh acts.
>
> What is the limit of such obedience? At many points we attempted to establish a boundary. Cries from the victim were inserted; they were not effective enough. The victim claimed heart trouble; subjects still shocked him on command. The victim pleaded that he be let free and his answers no longer registered on the signal box: subjects continued to shock him. At the outset, we had not conceived that such drastic procedures would be needed to generate disobedience, and each step was added only as the ineffectiveness of the earlier techniques became clear. The final effort to establish a limit was the Touch-Proximity condition. But the very first subject in this condition subdued the victim on command, and proceeded to the highest shock level. A quarter of the subjects in this condition performed similarly.
>
> The results, as seen and felt in the laboratory, are to this author disturbing. They raise the possibility that human nature, or more specifically the kind of character produced in American society, cannot be counted on to insulate its citizens from brutality and inhumane treatment at the direction of malevolent authority. A substantial proportion of people do what they are told to do, irrespective of the content of the act and without limitations of conscience, so long as they perceive that the command comes from a legitimate authority. If, in this study, an anonymous experimenter could successfully command adults to subdue a 50-year old man and force on him painful electric shocks against his protests, one can only wonder what government, with its vastly greater authority and prestige, can command of its citizenry [pp. 261–262].

Lest Milgram's research be considered idiosyncratic to a laboratory investigation, recent events deriving from the American experience in Vietnam bolster the cause of Milgram's concerns. The trial of Lieutenant Calley has caused extended debate regarding responsibility for horrendous acts committed in the service of legitimate authority. Calley, the officer in charge of an advancing platoon, gave orders for his men to kill each and every person in the village of My Lai, including infants, the elderly, and men and women alike.

Details of atrocities in Vietnam have been common enough that the My Lai slaughter might never have gained much attention. However, the very "atmosphere surrounding the massacre" with the resulting questions pertaining to personal responsibility generated a ground swell of public indignation. Some were outraged at the barbarism per se and of the denial of responsibility by different military officials. Others expressed indignation that a comparatively low ranking officer should have been assigned the blame which in their view should be shared by the entire leadership that created such a callous atmosphere. Still others resented the fact that a man in the service be held accountable at all since they did not see him as acting on his own volition.

Kelman and Lawrence (1972) conducted a survey in which a sample of Americans were questioned about the Calley trial and other similar instances in which responsibility for personal actions had been raised in the past; of 989 respondents only 34% approved of Calley's having been brought to trial. Most pertinent to the thesis to be presented in this chapter is the finding that those who believed that an individual soldier bears responsibility for his actions showed marked and consistent differences in their general beliefs about responsibility from those who believed that a soldier should not be held accountable for his actions.

Kelman and Lawrence interpreted responsibility attribution differences in terms of the manner in which individuals feel integrated into their national group. Those who denied responsibility were described as normatively integrated persons:

> An individual who is normatively integrated is bound to the system by virtue of the fact that he accepts the system's right to set the behavior of its members within a prescribed domain. Here we are dealing, one might say, with legitimacy in its pure form, in which the question of personal values and roles has become irrelevant. Acceptance of the system's right to unquestioning obedience may be based on a commitment to the state as a sacred object in its own right, or on a commitment to the necessity of law and order as a guarantor of equitable procedures. The normatively integrated member regards compliance with the system as a highly proper and valued orientation. When he is faced with demands to support the system he is likely to comply without question, since he regards it as his obligation to do so [Kelman & Lawrence, 1972, pp. 204–205].

Kelman and Lawrence note the similarity of the normatively integrated individual with deCharms' characterization of the "pawn" (deCharms, 1968). The pawn is said to perceive his actions as under the control of authorities, and in relinquishing control to these authorities, the pawn is said to disavow responsibility for his nation's policies.

In contrast, the person who views Calley and others like him as responsible for their actions is said to be characterized by ideological or role integration. Ideological integration occurs when the person "is bound to the system by virtue of the fact that he shares—at an internalized level—some of the basic values on which the system is established [Kelman & Lawrence, 1972, p. 20]." Such a person is said to comply with system demands when he sees those demands as consistent with the underlying values of the system to which he is committed. Role integration occurs when an individual is bound to the system through the roles he is engaged in and compliance depends on the extent to which the relevant role has been made salient by the situation at hand:

> In any event, whether their integration into the national system is primarily at the level of values or at the level of roles, most responsibility accepting respondents are probably socialized and integrated in a way that gives them a greater sense of ownership of the system. Their demographic characteristics, i.e., their generally higher levels of socio-economic status and education, are consistent with this orientation. Realistically, their educational background and the occupations to which it admits them provide them greater opportunity to become active agents in the society and a greater sense of

participation in shaping its policies and running its affairs. Even when they do not feel quite free to question national policies in terms of underlying values and to give or withhold their support on that basis, they are less inclined to the view held by most responsibility denying respondents that these policies are beyond question and that their personal values are entirely irrelevant to them [Kelman & Lawrence, 1972, pp. 207–208].

In their discussion concerning the perception of responsibility, Kelman and Lawrence accentuate one element that would seem to be a primary difference between those who accept and those who deny responsibility, that is, a skeptical or questioning attitude:

In the Milgram experiments not all of the subjects remained submissive to the experimenter; at My Lai not all of the men followed Calley's orders to shoot the captured civilians. Those who resist in these situations have somehow managed to maintain the framework of personal causation that applies in "normal" situations. Perhaps they have never made the radical shift in perception of the situation described by Milgram, or perhaps they are better able to differentiate between legitimate and illegitimate demands and feel freer to make choices on that basis [Kelman & Lawrence, 1972, p. 181].

Thus, resistance to orders and the acceptance of responsibility when one is compliant to them is said to derive from the person's maintenance of a framework of personal causation and the ability to differentiate or assess the quality of demands made upon him. As Kelman has suggested, the dimension of actor–pawn may well define those who can resist orders and accept responsibility from those who are more compliant but responsibility-eschewing individuals. Likewise, it will be asserted, the locus of control dimension describes the manner in which one commonly ascribes responsibility for one's experiences. Persons who perceive themselves as the active determiners of their fates should more readily accept responsibility for their outcomes and, therefore, should be more discriminating about what they will and will not do in obeisance to others.

We will propose in this and the following chapter that the maintenance of an internal control orientation is a bulwark against unquestioning submission to authority. The data in support of this contention can be viewed as inconclusive, and perhaps inadequate. Nevertheless, the consistency of results even when not anticipated allow us to step out on a limb in addressing one of today's major problems, the maintenance of a moral order among men.

EMPIRICAL RESEARCH

Conformity and Compliance Research

The first investigation linking locus of control to influence resistance was conducted by Odell (1959) who found a significant relationship ($r = .33, p < .05$) between a prototype of Rotter's I–E scale and Barron's Independence of Judgment scale. Externals showed greater tendencies to conform on Barron's scale. (The latter test had originally been developed as a correlate of conformity in Asch-type situations.)

A second study utilizing an Asch-like conformity situation was conducted by Crowne and Liverant (1963). These investigators assessed the confidence statements of students made during two conformity tasks. In one, the subjects were asked to rate how certain they felt about their own judgment, from 0 (no confidence) to 10 (very high confidence), immediately after making each judgment. The task required subjects to distinguish, on each of 20 trials, the larger of two groups of dots that had been presented tachistoscopically for a 1-second interval.

As is customary in Asch conformity tasks, the subject stated most of his decisions after his supposed peers had publicly stated their judgments. However, the peers were collaborators of the experimenter whose judgments were prearranged in such a manner that they often were in complete agreement about many choices that were obviously wrong. The subject, therefore, had to be willing to deviate from a consensus if he was to be true to his own senses.

The second manner of presenting this task substituted the betting of money for the simple statements of confidence. Subjects were given $10 in quarters. On each of 20 trials the subject could bet nothing, 25¢, or 50¢ on each of his judgments. Subjects had been instructed that they could keep a portion of their winnings so that there was some value in making accurate assessments.

Differences between statements of confidence by internal and external control-oriented subjects, as measured on the 60-item prototype of Rotter's I—E Scale, were in the expected direction but not pronounced. When betting was substituted for the verbal expression of confidence, however, externals were found to conform more than internals ($t = 2.35$, $p < .05$). Of even greater interest, externals bet less when their judgments were independent of their peers than they did on trials on which they had yielded to the consensus ($t = 2.68, p < .02$); that is, when externals conformed to their peer's decision they were willing to bet larger amounts of money on the correctness of those decisions than they were when they made less conforming or independent judgments. The greatest differences between internals and externals in amounts bet occurred on those trials during which subjects made independent judgments, that is, on those trials when they responded before their peers had expressed their judgments.

The Crowne and Liverant study indicates that when the stakes of success are of some value to the individual, persons characterized as internals are more trusting of their own judgments than are externals. The difference between the groups was apparently due to the fact that externals have more confidence in the consensual judgments of others than they do in their own independent judgments.

At about the same time that Crowne and Liverant were examining the correlates of conformity, Pearl Gore (1962) was investigating personal characteristics that contributed to compliance with the experimenter's bias during

psychological experimentation. Gore administered a set of Thematic Apperception Test pictures with the stated purpose of learning which cards provoked
longer stories from subjects. During the test administration through smiles and
vocal intonations, she attempted to manipulate subjects into producing longer
stories to selected pictures. In one condition, in which Gore offered no hints or
suggestions as to what was expected, no differences were obtained between
internals and externals. Likewise, when the experimenter explicitly indicated
which pictures she thought were most likely to elicit longer stories no differences were obtained. However, when the experimenter smiled and dropped such
choice comments as "Now let's see what you can do with *this* one," internals
produced significantly shorter stories than did externals.

Gore's investigation suggests that internals are not more resistant to the
experimenter's bias in general than are externals, but that they resist a certain
kind of influence technique. Some writers have termed this technique subtle
manipulation, as opposed to explicit suggestion. However, it is also possible to
construe the technique difference in the language of interpersonal intentions.

If an experimenter conveys his expectations to a subject he may, in effect, be
perceived as sharing his hypothesis with the subject, thus inviting the subject to
join him in a curious objective attitude toward the criterion behaviors. This
shared attitude would be one that expresses respect for the subject as a person
rather than as an object to be manipulated and influenced. In contrast, the
"subtle coercion" described above pits experimenter against subject and suggests
that the subject will be affected as if a passive pawn by the characteristics of the
TAT card. The experimenter is, in effect, saying "you won't be able to resist this
one—I know," suggesting that the experimenter knows something that the
subject does not and is therefore in a position to more easily predict the
subject's behavior than is the subject.

It will be our contention as we review other relevant investigations that the
internal resists being placed in positions where he is the pawn who is "put
down" so to speak, by the assumed knowledge of others. If the expert or
experimenter, on the other hand, offers a more collegial atmosphere the internal
should not then be as resistant to influence as when he perceives himself as the
target of manipulation.

Two studies are immediately relevant to this conjecture. Strickland (1970)
found that internals were more likely than externals to deny having been
influenced during a verbal conditioning experiment. In this task subjects were
reinforced, that is, the experimenter nodded her head and murmured "umm-
humm" after the subject answered with verb responses. The subject's job was to
select a word from a group of four words that went best with some common
noun. The effect of reinforcement was expected to increase the choice of a verb
as the proper response *unbeknownst* to the subject; that is, the subject was to be
"fooled" by the experimenter into doing the latter's bidding without his conscious consent. At the end of the task subjects were interviewed to ascertain
whether they had become aware of the reinforcement contingency. Internals and

externals did not differ with regard to their being aware of the experimenter's behavior. However, among those subjects who had become "aware" of the contingencies, externality was associated with acceptance of the experimenter's influence. "Subjects characterized as internal tended to deny the influence of the experimenter and appeared to follow their own inclinations in regard to giving the correct response [Strickland, 1970, p. 376]."

In other words, among these subjects in Strickland's experiment who recognized that the experimenter was doing something deliberate to manipulate them, and who also denied being influenced, the more internal individuals proved to be less conditionable. On the other hand, toward the end of the extinction trials, after the experimenter had ceased her cajoling manipulative responses, the internal-aware subjects showed a significant increase in verb responses. In essence, this small group of subjects seemed to say—"I know what you're trying to do to me and I'll show you you can't make a fool out of me," in essence a reassertion of "I, the actor," and a denial of "me—the object of manipulation."

Strickland's findings are mirrored fairly closely in another verbal conditioning experiment conducted by Getter (1966). This investigator found that the most responsive "conditioners" were his most external subjects. The more internal subjects, as in Strickland's experiment, behaved in a seemingly paradoxical fashion, producing the conditioned response (words ending in *ion*) largely during extinction trials, after the experimenter had ceased emitting reinforcements.

In these three studies, then, internals are found to behave in a somewhat oppositional manner, doing the reverse of what the experimenter would trick them into doing. Only when Gore explicitly instructed her subjects as to what she anticipated did the difference in compliance between internals and externals diminish.

Other researchers have reported findings that help us avoid falling into the trap of assuming that internality is simply a euphemism for stubbornness. As in Gore's study, Ritchie and Phares (1969) did not find that internals were consistent resisters to influence, while externals were always likely to succumb. Internals shifted their opinions in the direction of an influence message as did externals in this experiment. However, externals were found to shift their views most when the influencial arguments were attributed to a prestigious government official. Internals did not differ in their response as a function of the status of the source. When the arguments regarding government budgets were attributed to a college sophomore they were no less effective with internals than they were when attributed to the then U.S. Secretary of the Treasury (Douglas Dillon). By contrast, as can be seen in Figure 1, externals exhibited more attitude shift with the prestigious attribution and less with the low status attribution than internals who seemed responsive to the arguments regardless of the assumed source.

As further evidence that internals are not always more resistant to influence than externals, James, Woodruff, and Werner (1965) found that more internal males quit smoking for a specified length of time than did external males after

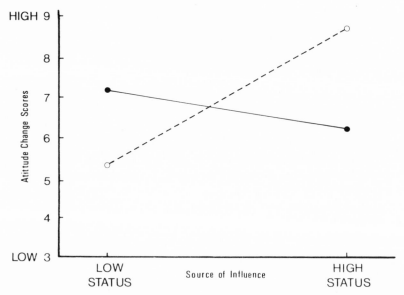

FIGURE 1   Mean attitude change scores as a function of locus of control and status of influence source. (●) Internal; (○) external. (Adapted from Ritchie & Phares, 1969, page 438.)

hearing the United States Public Health Service Surgeon General's report concerning the link between cancer with cigarette smoking. Likewise, E. S. Platt (1969) reported more success at influencing the smoking behavior of internals than of externals. Platt used role-playing procedures in which subjects acted out the roles of physician and patient during a medical examination. The examination concluded with a diagnosis of cancer and patients had to actively think about the arrangements necessitated by hospitalization and imminent death. Platt found that the greatest changes in smoking behavior occurred among individuals who were both internal and believers in the link between smoking and illness.

These investigations suggest that internals are not simply resistant to influence, but are discriminating about what influences they will accept. Authority per se has little direct effect upon their readiness to accept a point of view. However, when internals accept the information, as in the Ritchie and Phares investigation and in the smoking studies discussed above, they do change behaviorally in response to influence. In each of these studies where internals do shift in their attitudes or behaviors there is less likelihood that subjects view themselves as manipulated objects. The prestige condition in the Ritchie and Phares study is the only instance in which "oughts" or pressures for conformity are evident, and in that condition internals were less influenced than were externals. The smoking studies had as vehicles of influence nonpersonally directed public information and role playing in which subjects took an active part in creating their responses.

In neither case was external pressure brought to bear in a way as to curtail the decision-making process. The subject had to arrive at his own conclusions and it is this opportunity, perhaps, that is the essential precondition if internals are to be influenced.

Two more recently reported investigations expressly concerned with locus of control and responsiveness to influence underscore the reliability of the previously reported findings (Biondo & MacDonald, 1971; Doctor, 1971). Biondo and MacDonald attempted to manipulate the "level of subtlety" in their study so as to examine previous contentions that internals are more markedly resistant to subtle than to obvious influence attempts. The curves in Figure 2 suggest that, regardless of the style of influence, externals were more compliant than internals. In addition, the latter appeared to be even more oppositional in the presence of obvious coercion than they were with subtle influence. These findings seem to refute the conjectures offered previously. If one examines the methodology employed in Biondo and MacDonald's study, however, one can as easily conclude that subjects perceived themselves as objects of manipulation in both the "low" and "high" influence conditions.

Three different groups of subjects were asked to express on paper their evaluation of the marking system at their university. Immediately after completing their evaluations and passing them forward, another person, introduced as a graduate student in education, addressed the class and explained that he was

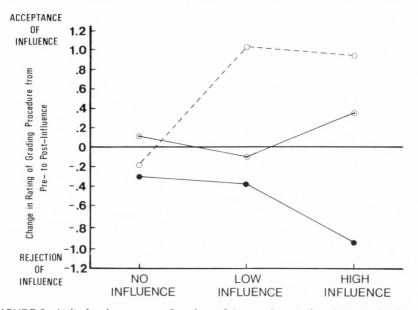

FIGURE 2 Attitude change as a function of locus of control and level of influence manipulation. (●) Internal; (○) external; (◉) middle. (Adapted from Biondo & MacDonald, 1971, page 414; © 1971 Duke University Press.)

conducting research upon grading systems. The experimenter then passed out mimeographed sheets containing information about the grading procedure which had just been rated. One group received no additional information from that which had been provided earlier. The low influence communication ended with the statement that "This grading procedure has been widely accepted at other universities and appears to be one of the best ever used." The high influence communication ended with "Taking everything into consideration, it is obvious that this is a very good procedure. I don't see how you have any choice but to rate this procedure highly."

Following these influence attempts subjects were asked to rerate the grading procedure. Although Biondo and MacDonald (1971) reported that there were no expressions of suspicion obtained, the very procedures described here which involved a sequence of rating—information—rerating seem to this writer to be an obvious attempt to assess suggestability without admitting as much. In essence, subjects are given no chance to debate or defend their position, little time to actively rethink their reasons for previous judgments, and are placed in a position in which if they do shift their judgments they must view themselves as easily manipulable, deferent to "expert opinion," and perhaps unstable. To so rapidly alter one's opinion must be perceivable by subjects as the experience of being influenced. As such internals, if our thesis regarding aversion to passive manipulation is correct, should resist such attempts at influence whether the statements are short (low influence) or long and somewhat embellished (high influence).

The other study (Doctor, 1971) has essentially replicated the verbal conditioning results obtained by Strickland and Getter. The one difference was that awareness of contingencies was not associated with a negative response on the part of internals. Rather, externals who were aware of the contingencies exhibited the greatest degree of conditioning, which accounted for most of the differences obtained between internal and external subjects. In other words, external subjects, on becoming aware of the experimenter's purpose, willingly submitted to his machinations, whereas internals showed no clear disposition to cooperate or resist.

Resistance to Experimenter's Influence

In addition to those investigations which have been specifically designed to assess compliance, a series of studies conducted by this author and his colleagues (Lefcourt, 1967; Lefcourt, Lewis, & Silverman, 1968; Lefcourt & Siegel, 1970a; Lefcourt & Wine, 1969) revealed that internals were unresponsive to the experimenter's instructions, suggestions, and manipulations, whereas externals readily capitulated, behaving, almost to a man, in accord with task directions. In one study (Lefcourt, 1967), the achievement-oriented behaviors of internals and externals were compared when instructions varied the degree to which achievement was stressed. Achievement-oriented patterns of performance were obtained from 91% of the external subjects when task directions emphasized the challenging nature of the task. When the task (Rotter's Level of Aspiration Board; Rotter, 1954) was administered without such instructions only 18% of external subjects responded in an achievement-oriented fashion. In contrast, samples of internal subjects exhibited little change in behavior as a function of the different instructions.

In another study using the level of aspiration technique, Lefcourt, Lewis, and Silverman (1968) initially found no performance differences between internals and externals in response to skill versus chance directions which were given before the task. However, at the end of the experiment, when subjects were asked how they had actually perceived the task, externals were found to have been more accepting than internals of instructions which suggested that this truly skill-demanding task was chance determined. On the other hand, externals had been somewhat less likely than internals to have accepted the more veridical skill instructions. It was concluded, therefore, that internals were less compliant than externals in response to directions which challenged their own more probable interpretations.

A similar finding in regard to the way in which externals respond to experimenter's directions was reported by Lefcourt and Wine (1969). Part of this experiment had concerned the effect of the experimenter's stated purposes on subjects' attentiveness. The number of observations of things unique to an experimental room were recalled by internals and externals in two conditions: one in which the purpose of the experiment was not defined for the subjects, and another in which the experiment was described as focusing upon attention. All subjects completed a series of unrelated tasks while in the experimental room. The number of observations made of items in that room was obtained after subjects had been seated in a second room. Unique to the experimental room were several "planted items": an erotic Shakespearian sonnet on the blackboard, signs advertising various events, and so on. The number of unique stimuli recalled was much greater overall when the experiment was described as pertaining to attention. Moreover, the external group exhibited the greatest shift in the number of observations made. Internals had observed somewhat more of the unique items than externals in the noninstructed condition, though this difference was not statistically different. With instructions focusing upon attention, however, externals recalled significantly more unique items than did internals, and more than their external counterparts in the uninstructed condition. Internals, on the other hand, barely changed at all with the different instructions.

In this experiment then, the directions which simply alerted subjects as to what processes the experimenter was interested in had the effect of altering externals' perceptual behavior. Since subjects hadn't been aware that the immediate room was to be the target of their attentions, it was as if externals began to use their eyes when they knew that the experiment was concerned with attention. In contrast, internals seemed unresponsive to the experimenter's interests.

In these and other studies then, evidence has been found which indicates that persons holding an internal locus of control withstand pressures directing them to behave in certain circumscribed manners. This is not true in all instances, although the exceptions to this generalization are revealing. Internals do yield to pressures, but not to the same pressures as externals. When acted upon as an object of experimentation, internals become almost playfully negativistic, as in the verbal conditioning investigations. However, internals do respond positively to reasoned arguments, regardless of the status of the source, readily respond to directives that seem congruent with their own perceptions, and shift in attitudes and behavior when engaged in role playing, which allows more active participation and self-direction. Externals appear to be responsive to the status of the influencer and more ready to accept the suggestions and directions of an experimenter.

Morally Relevant Behavior

At this point we may now return to the reason for our initial interest in the resistance to influence. It would seem from the studies reported above that an internal locus of control is associated with a tendency to be circumspect in the face of pressure to yield to influence. However, as of yet, there have been no investigations linking locus of control with the kinds of influence resistance that have been investigated by Milgram. Nor has locus of control been implicated in studies of complicity in evil or immoral acts. Internals are more likely to resist some forms of social pressure than are externals. Whether such resistance would persist against heightened pressure or increased inducements to acquiesce are questions requiring further empirical test.

There are a few investigations, however, which do approach the issue of moral responsibility and locus of control. Johnson, Ackerman, Frank, and Fionda (1968) have investigated the relationship between locus of control and the resistance to temptation as part of a project concerned with moral development and personal adjustment. These authors used a "complete a story" device in which the hero of an incomplete story experienced social pressure directing him toward the violation of some social norm (drug use, illicit sex, etc.). Subjects completed stories in which the hero was either at the point of decision or had to face the consequences of his acts. Among male undergraduates, Johnson *et al.* (1968) found that the more internal the subject, as measured by the I—E scale, the more likely was he to complete stories in which the hero resisted pressure. And, in those stories in which the transgression had already occurred, internals were more likely than externals to have the hero acknowledge guilt about his having yielded to pressure.

While the reliability and validity of these data can be subjected to question, taken together with the research noted above the findings are intriguing. Johnson's results indicate that the relationship between locus of control and the resistance to influence can be extended to moral decision making. A recent study by Charles Johnson and John Gormly (1972) provides some support for the link between locus of control and resistance to temptation. When fifth-grade boys and girls were classified as cheaters and noncheaters on the basis of a behavioral test, a significant difference in scores on Crandall's IAR scale was obtained. Female pupils who had cheated were more external than their noncheating peers. Male pupils had results in the same direction though the differences between cheaters and noncheaters was decidedly less extreme than among the females.

In another study with pertinence to morality, Midlarski (1971) found that internals were more likely to help another individual than were externals despite the fact that they were penalized for doing so. In this investigation, subjects found themselves in a position where it was possible to help a fellow subject (actually a confederate of the experimenter) to finish a task of sorting small objects into separate groups. As each item was listed from a grid, a moderate-level electric shock was received by the subject. The confederate performed at a slower rate than the subject such that the latter had the option of helping the confederate after finishing his part of the task. A fairly strong correlation ($r = .54, p < .001$) was obtained between helping and Midlarski's shortened form of the I–E scale.

The latter study indicates that internals are more tolerant of discomfort in doing what they consider to be correct than are externals. This finding matches that of Johnson et al. (1968) whose subjects had to make the choice between resisting pressures to commit immoral acts and suffering ostracism, loneliness, and so on. In the one case internals tolerated pain for actively doing what they considered correct, while in the other they expressed a willingness to risk social rejection for maintaining what they construed as proper behavior.

On last indication of the relevance of locus of control to morality derives from a low magnitude, but significant correlation ($r = .16, p < .05, N = 227$) between the I–E scale and the Mach V scale (Miller & Minton, 1969). The latter scale assesses the espousal of Machiavellian attitudes including the readiness to be opportunistic and to be guided by indiscriminate cynicism and suspicion. As the correlation indicates, the more external persons tend to agree with Machiavellian positions more than do internal individuals.

CONCLUSIONS

Though the research discussed in the previous section lacks the dramatic quality so obvious in Milgram's research, it lends support to the contention advanced earlier that internal locus of control can operate as a bulwark against the unquestioning submission to authority. Insofar as the researcher is a legitimate authority and defiance against his often demeaning requirements can be taken as the readiness to resist authoritarian dictates, there is some encouragement for discussing locus of control as pertinent to the Eichmann and Calley cases. If these individuals had remained able to question the commands and legitimacy of their superior officers they might not have been the infamous "collaborators" that they did in fact become; and, if they had perceived themselves as responsible actors rather than as externally controlled pawns, they might have been more questioning and consequently more resistant to the dictates and persuasions of others.

If our hypothesis is correct, when a person believes that he is the responsible agent or source of his own life's fortunes, he will resist influence attempts which aim to bypass his own sense of moral justice, and will only respond to those appeals that address themselves to his own beliefs and values. In effect, the internal will not, like Eichmann, experience the "Pontius Pilate" feeling, a surrender of a sense of responsibility when one succumbs along with others to manipulation.

5
Locus of Control
and Cognitive Activity

INTRODUCTION

INTRODUCTION

In the previous chapter, Kelman and Lawrence (1972) explained the differences between persons who indiscriminately capitulate and those who selectively respond to the demands of authority as follows: "Those who resist . . . have somehow managed to maintain the framework of personal causation that applies in normal situations. Perhaps they have never made the radical shift in perception of the situation . . . or perhaps they are better able to differentiate between legitimate and illegitimate demands [p. 181]." Inherent in this explanation is the assumption of cognitive activity that facilitates making fine discriminations. Individuals must maintain their sense of personal causation, according to Kelman and Lawrence, despite the onslaught of bewildering situations that tempt them to surrender their sense of personal control and responsibility.

Writers throughout the ages have depicted man's struggle to maintain his sense of personal responsibility in the face of pressures that threaten to destroy his very individuality. *The Book of Job,* Kafka's *The Trial,* Koestler's *Darkness at Noon,* and Heller's *Catch 22* present man as he is twisted by God, fate, the Devil, and relentless, omnipotent social systems. With such forces arrayed against him it is perplexing that man attempts to and often succeeds in remaining stable in his "framework of personal causation."

It will be contended here that the degree to which man is capable of questioning his own assumptions, is in the habit of deliberating over his options, and is attentive to information relevant to his decision-making, to that degree will he be less assailable by forces directed against his integrity. If Eichmann had been more able to ponder the ramifications of Nazi requests for complicity, perhaps he would have looked more askance at the enthusiasm exhibited by his fellow bureaucrats for participating in the execution of the "Final Solution." In other words, a man can be led more easily to surrender his moral precepts when he has, through lack of thought, failed to discern the portents inherent in his decisions.

The investigations that are reviewed in this chapter indicate that locus of control is a correlate of the kinds of cognitive activity which should facilitate the maintenance of personal causation. If an individual wishes to know about himself, his capabilities, his limits, and his potentialities, then he should seek out situations in which it is possible to test his mettle. Persons who customarily do so are referred to as "open to experience," "nondefensive" and "self-actualizing" in various theoretical frameworks. For the present purposes, such individuals who readily assimilate information about themselves will be described as having an internal locus of control.

The idea that locus of control is related to cognitive activity appeals to common sense. Persons holding internal control expectancies should be more cautious and calculating about their choices, involvements, and personal entanglements than are individuals with external control orientations. Otherwise, the probability of internals being able to regulate their experiences would be lessened which, in turn, would diminish the degree to which they are able to remain actors rather than pawns of fate. Such self-direction should entail more active cognitive processing of information relevant to the attainment of valued ends and should be reflected in the types of strategies that characterize an individual.

INFORMATION ASSIMILATION

The first study linking locus of control and cognitive activity was conducted by Seeman and Evans (1962). They used a 12-item measure of powerlessness, derived from Rotter's I—E scale, to predict knowledge about a disease among sufferers of that disease. External-oriented tubercular patients were found to have less knowledge about tuberculosis than internal tubercular patients, the correlation between powerlessness and knowledge being $r = .31$, $p < .01$. In addition, when the staff of the sanitorium was asked to estimate their patients' knowledge about tuberculosis, internals were rated as more knowledgeable than externals.

The results of this study support the assertion that internals avail themselves of information, even if it has negative connotations for themselves, more than do externals. It has been assumed that this difference derives from the fact that internals believe that they can act in their own behalf and therefore require more information, while externals more readily accept dependency on more competent others and thus have less need of information.

Seeman (1963) tested these assertions in another institutional setting with a measure of active information gathering that had definite functional utility. He presented reformatory inmates with three types of information varying in utility, and six weeks later assessed the subjects' retention of that information. Of the three types of information, only one type, related to parole attainment, was of instrumental relevance to the reformatory inmates. Examples of parole information were as follows:

1. About 85% of successful parolees have taken part in some kind of voluntary education program during their prison term.
2. In 60% of the cases where parole is delayed beyond the date set by the board, difficulty in arranging employment is the main reason for the delay.

The other types of information were "incidental knowledge about the reformatory," examples of which are:

1. It costs about $5 per day to keep an inmate at this reformatory; and
2. A survey at the reformatory showed that 65% of the men had no disciplinary actions on their record;

and "knowledge of long range, distant opportunities," examples of which are:

1. It has been estimated that the demand for unskilled workers will continue to decline in the 1960s, with the loss of about 2500 such jobs per year; and
2. A sizable increase in federal funds for vocational training of young adults who have a prison record is predicted by 1965.

Seeman predicted that powerlessness would be related to more accurate recall of parole information, but not to the recall of less goal relevant types of knowledge. The results supported Seeman's hypothesis: internals recalled more parole relevant information than externals ($r = -.23$, $p < .05$) while incidental reformatory information ($r = -.16$, n.s.) and long-range opportunity information ($r = -.09$, n.s.) were not significantly related to the powerlessness measure. Furthermore, when Seeman differentiated among inmates on the basis of conduct reports, for those who were more or less likely to be hoping for parole, the relationship between powerlessness and parole knowledge increased considerably ($r = -.40$, $p < .05$).

Seeman's results led him to conclude that an individual's sense of powerlessness governs his attention and acquisition processes. Since Seeman reported his results, several investigators have published research findings that bear further upon the hypothesized relationship between locus of control and cognitive activity. In one such study, Davis and Phares (1967) gave subjects the task of influencing another person regarding his attitudes toward the Vietnam war. The subjects were led to believe that the experimenters had a file of data available about each individual who was to be the target of influence. The primary measure was the number of questions that each subject asked about the specific individual whom they were to influence.

The investigators anticipated that internals would generally seek more information than would externals in order to improve their likelihood of being effective. In addition, they examined the effects of instructions given to the subjects: one group received "skill directions," which stressed ability as the major determinant of success; a second group received "luck directions," which emphasized the fortuitous combination of the personality characteristics of influencer and influencee as primary causal factors of influence success; a third group was given no specific instruction regarding the causes of success or failure at influence.

Internals asked more questions than externals when skill and no instructions were given, as may be noted in Figure 3. When the directions stressed luck as the causal factor, internals and externals did not differ from each other. However, it is notable that externals asked more and internals asked fewer questions in the chance condition than the equivalent groups in the skill and no instruction conditions.

These results would seem to indicate that internals are more likely to engage in the preliminary steps of data gathering than are externals when information seeking seems pertinent to outcome determination. In turn, such information seeking should increase the probability that internals will more often succeed in skill-demanding tasks than their external counterparts.

In another study reported by Phares (1968), internals and externals were compared in their use of information for decision making. All subjects learned ten bits of information about each of four men until they were able to recall that information without error. A week later, the subjects were asked to guess who of eight girls, and which of ten occupations, were best suited to each of the four men. Financial rewards were offered for correct matchings, and subjects

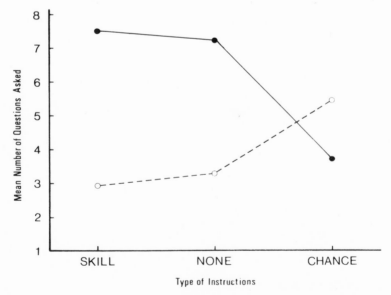

FIGURE 3   Mean number of questions asked as a function of locus of control and the type of instructions. (●) Internal; (○) external. (Adapted from Davis & Phares, 1967.)

were asked to list their reasons for their matches. The criterion measure was the number of reasons provided for each match. Internals were found to give more reasons, more than 50% more, than externals. When only correct reasons were counted the difference increased dramatically: internals gave more than three times as many correct reasons as externals for justifying their social and occupational matchings. These differences led Phares to conclude that internals make better use of information than externals despite the fact that both might have equivalent funds of information.

ATTENTION

Another cognitive function that has been examined in locus of control research is that of attention, that is, the ways in which individuals focus upon cues of relevance for goal attainments. Lefcourt and Wine (1969) observed subjects as they attended to another person with whom they were trying to become familiar. The subject's task was to interview two of the experimenter's assistants, with the object of finding out as much as possible about each of them. Ultimately they were to write personality descriptions of the assistants as part of a final examination for a personality course. One assistant behaved in an appropriately pleasant manner, engaged in eye contact, and often smiled and exchanged pleasantries with the subject. The second assistant studiously avoided eye contact with the subjects and behaved in a restrained aloof manner. It was hypothesized that the uncertainty caused by the behavior of the second assistant would arouse more curiosity and attentiveness among internal than among external subjects.

In Figure 4, the frequency of looking at the facial area of the students is plotted as a function of the assistant's behavior and the subject's locus of control. The curves illustrate that internals attended to the assistant's face more often than externals when the assistant behaved in the more puzzling avoidant fashion. Externals, on the other hand, looked somewhat more at the conventionally behaving assistant. These differences produced a significant interaction ($F = 6.13$, $p < .02$, 1/16), indicating that locus of control and the assistant's actions jointly affected the subject's looking behavior.

When the number of behavioral observations reported in final papers was counted, internals had made more observations than externals about both the conventional and puzzling assistants ($F = 4.33$, $p < .05$, 1/20). Internals had recorded more observations ($M = 6.09$) of the quizzical than of the conventional ($M = 5.00$) assistants, though the differences were not significant. In addition, attention to the face of the quizzical assistant was positively related to the number of observations reported ($r = .68$, $p < .01$).

Lefcourt and Wine concluded that internal subjects were more likely to attend to cues which help to resolve uncertainties. The more conventional assistant

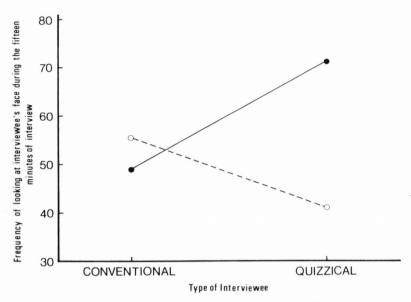

FIGURE 4   Frequency of looking at the interviewee's face as a function of locus of control and interviewee behavior. (●) Internal; (○) external. (Adapted from Lefcourt & Wine, 1969.)

offered less challenge to subjects because he was predictable. The aloof assistant, by contrast, behaved less predictably and, thus, created an interpretation challenge for the subject: "Was the assistant inordinately shy? Was the subject repulsive to the assistant? Or, was this a contrived situation and a test of his own ability to draw out the assistant?" Since the task at hand was to make sense of the assistant's verbal and nonverbal actions, the subject had to be vigilant if he was to be able to resolve the uncertainties that the aloof assistant aroused.

Three different studies beginning with one by Rotter and Mulry (1965) have indicated that internals devote more attention to decisions about skill-related matters than do externals. Rotter and Mulry used a very difficult matching task similar to those described in Chapter 3 (James, 1957; Phares, 1957). Half of the subjects were instructed that the perceptual acuity required of them for successful matching was a special skill; the other half were instructed that success was more a matter of chance. In the actual matching task all subjects received the same sequence of "rights" and "wrongs." The major findings from this study are illustrated in Figure 5, which makes it clear that internals spent more time deliberating about their decisions than did externals when the task was thought to be skill demanding. Externals were somewhat more deliberate with chance directions though this difference was not statistically significant. The major factor that produced the interaction ($F = 6.67$, $p < .025$, 1/116) was the increased length of time taken by internals to make decisions in the skill condition.

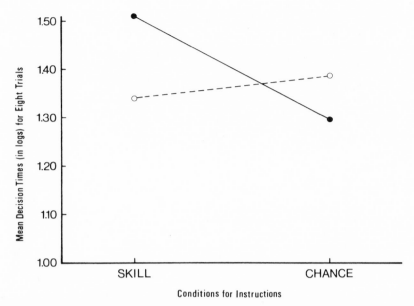

FIGURE 5   Mean decision times as a function of locus of control and skill—chance instructions. (●) Internal; (○) external. (Adapted from Rotter & Mulry, 1965.)

These results reflecting upon the different approaches by internal and external individuals to skill versus chance tasks have been replicated with some variation in two later studies. Julian and Katz (1968) reported that internals required more time to make decisions as the difficulty of decision making increased. Externals did not vary as much with the difficulty level of the tasks, behaving as if there were no differences between simple and difficult choices. Another similar experiment (Lefcourt, Lewis, & Silverman, 1968) essentially replicated the Rotter and Mulry study with some embellishments. These investigators used the Level of Aspiration Board, described previously, and tried to alter subjects' expectancies concerning the skill—chance nature of the task through instructions. Initially, no differences between internal and external subjects were obtained with respect to decision times. In retrospect, these investigators suspected that some of their subjects might have regarded chance instructions as specious since the Level of Aspiration task obviously required some skill. Therefore, subjects' written reports were examined to see whether they believed the instructions they had received. It was found that externals had accepted the experimenter's instructions more readily than had internals, especially when those directions had emphasized chance determination. Decision-time data were then reanalyzed using the subjects' own judgments as to the skill and chance determination of the task outcomes. With subjects' perceptions used to form groupings the results obtained were as illustrated in Figure 6. On each observed characteristic, internals were found to be more cognitively engaged when they perceived the Level of

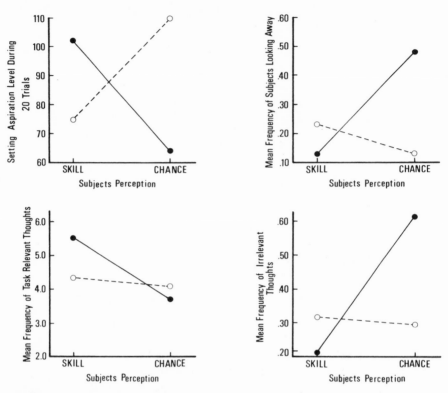

FIGURE 6   Decision times, inattentiveness, task relevant, and task irrelevant thoughts as a function of locus of control and skill versus chance perception of task. (●) Internal; (○) external. (Adapted from Lefcourt, Lewis, & Silverman, 1968.)

Aspiration Board as a skill-demanding task. Internals who perceived the task as skill determined spent more time in decision making, were less likely to allow their attention to wander from the task, recalled having had more task relevant and less task irrelevant thoughts than did internals who believed that the task was chance-determined. The reverse tended to be true of externals, though it was only with regard to decision-making times that the magnitude of their change was sufficient to generate statistical significance.

In the aforementioned studies then, internals were found to show considerably more variability than externals. Their attentiveness, concern, and interest changed with the types of situations in which they were engaged. If the task offered a challenge to competence, then internals became more deliberate in their decision making during that task. Less skill-demanding tasks, on the other hand, elicited some carelessness and impulsivity from internals. Externals did not seem to draw such sharp distinctions about tasks as internals were wont to do. When task instructions did affect them, however, it was the more chance-determined task which elicited their greater attention and deliberation.

PSYCHOLOGICAL DIFFERENTIATION

Pursuing the concern with attention differences, Lefcourt and Siegel (1970a) examined the reaction time performances of internal and external subjects. While no differences were obtained in reaction time performance as a function of locus of control, these investigators (Lefcourt & Siegel, 1970b) did find some success in predicting performance with a measure of field dependence: field independent subjects had shorter reaction times (greater attentiveness) with self as opposed to experimenter-controlled stimulus onset during the reaction time procedure. More important for the present purposes, however, was the reasoning for the use of the field dependence variable in this experiment. Witkin and his colleagues (Witkin, Dyk, Faterson, Goodenough, & Karp, 1962) have expanded upon their earlier research with perceptual field dependence, describing a concept referred to as psychological differentiation. These writers speak of differentiation and identity in such terms that the overlap with the locus of control construct is apparent:

> With respect to relations with the surrounding field, a high level of differentiation implies clear separation of what is identified as external to the self. The self is experienced as having definite limits or boundaries. Segregation of the self helps make possible greater determination of functioning from within, as opposed to a more or less enforced reliance on external nurturance and support for maintenance typical of the relatively undifferential state [Witkin, Dyk, Faterson, Goodenough, & Karp, 1962, p. 10].

A relationship between I–E and differentiation, or at least a similar pattern of relationships for these variables with different criterion behaviors, could be anticipated on the basis of the apparent similarities between the locus of control and differentiation constructs.

Rotter (1966) had previously reported that no empirical relationship had been found between the I–E scale and the Gottschalk Figures Test (one measure of differentiation). Chance and Goldstein (1967), likewise, found a nonsignificant relationship between I–E and performance on the Embedded Figures Test. However, in the latter investigation, internals improved steadily from trial to trial as they progressed through the Embedded Figures Test. Deever (1968) and Lefcourt and Telegdi (1971) found no relationship between the rod and frame, block design, and embedded figures measures of differentiation, and the locus of control. Nevertheless, the constructs bear similarity in predicting assertiveness, the experiencing of oneself as a distinct source of causation, and the tendency to be self-reliant rather than acquiescent and conforming. That these variables might prove complementary to each other in predicting such criteria has received some support. Bax (1966) found significant correlations between I–E and a TAT measure of differentiation ($r = .37$, $p < .001$, $N = 96$) and lack of assertiveness ($r = .35$, $p < .001$, $N = 96$) in the expected directions.

Deever (1968) found that high differentiation and internal control were related to a greater reliance upon one's own reinforcement history than on group norms for the prediction of performance. In this latter experiment, subjects had to state how sure they were of success for each trial in which they were to compare the length of two similar lines. The procedure was a typical level of aspiration task with two additional elements: correctness feedback which was rigged beforehand; and subjects were provided with norms of peer performance for each trial.

Both differentiation, as measured by the Embedded Figures Test, and locus of control generated significant main effects. Internals and highly differentiated subjects expressed greater confidence of success and differed more from the provided norms than externals and less differentiated subjects. Despite similar relationships with the criterion measures, locus of control and differentiation were unrelated ($r = .02$).

Lefcourt and Telegdi (1971) used both locus of control and the rod and frame measurement of differentiation in an attempt to predict scores on measures reflecting cognitive activity. The dependent measures consisted of performance on Mednicks' Remote Associates Test, Barron's Human Movement Threshold Inkblot Test, and an incomplete sentences test. As predicted, internal–highly differentiated subjects surpassed all other subjects on each measure. However, external–low differentiated subjects proved to be second, and the incongruent groups (internal–low differentiated, external–high differentiated) were the poorest on each measure. In this study neither locus of control nor differentiation alone produced a significant main effect in the prediction of cognitive activity. Only in combination were significant results obtained.

RECENT CONFIRMATORY INVESTIGATIONS

Throughout most of the research described above, evidence has been presented that supports the assumed relationship between locus of control and cognitive activity. Whether the focus has been on attention, deliberation, inquisitiveness, or utilization of information, internals have more often been found to be active and alert individuals than have externals. Sometimes the differences between internals and externals have varied with the ostensible qualities of the task or have occurred only in combination with other measures. Nevertheless, an internal locus of control seems to be a sine qua non of being able to steer one's self more clearly and appropriately through the vagaries and confusions of different situations.

Two recent investigations have strongly reinforced the findings noted previously and underscore the linkage between locus of control and cognitive activity. Lefcourt, Gronnerud, and McDonald (1973) enlisted a sample of subjects for an experiment ostensibly concerned with verbal facility. The subjects were tested on a battery of measures requiring different verbal skills. Last among these devices was an orally administered word association test during which subjects had to respond to a series of verbal stimuli with the first word that entered their minds. The list of stimulus words, however, was no ordinary list, but contained a gradually increasing number of sexual double entendres. As can be seen in Table 1, the list began innocently enough until number 13, "rubber." Subsequently, every third stimulus word was another double entendre until Number 24. Afterward, the number of double entendres increased until, at the end of the list, all words were sexual double entendres.

On completion of the previous verbal tests subjects were, no doubt, expecting more of the same—more predictable, almost boring tests on the way to fulfilling their required stint as subjects. Beginning with the first double entendre, however, the seeds of doubt were planted. With the word "rubber" subjects were often bemused at their own private associations. Only rarely, however, were they apt to redefine the task at this point. But, after "bust" and then "snatch" the alert subject should have become suspicious, slowly concluding that the test was something other than what the experimenter had purported it to be. In short, where the initial covert associations were likely to be attributed to one's own predilections to think "dirty thoughts," the subjects should gradually have shifted their attribution of responsibility to the experimenter—"It's not me, but him—he's doing this deliberately for some undisclosed purpose."

While the test was being administered the time that was required for the subject to respond was closely monitored. The latency of a response given to a stimulus word in word association tests has classically been used as a measure of conflict. If a subject entertains two competing responses and must choose between them, then the time taken to respond will be substantially longer than when he has only one potential response to give to a particular stimulus. In this experiment, the delay in response time was used to indicate that subjects were

TABLE 1
Stimulus Words in the Double Entendre Word
Association List[a]

1 fly	11 light	21 sugar	31 measure	41 *HUMP*
2 face	12 work	22 *NUTS*	32 *BLOW*	42 *PET*
3 plant	13 *RUBBER*	23 cross	33 garden	43 *TOOL*
4 voice	14 health	24 *MAKE*	34 *COCK*	44 *SUCK*
5 earth	15 ocean	25 carpet	35 stove	45 *BANG*
6 miss	16 *BUST*	26 *CRACK*	36 *MOUNT*	46 *ASS*
7 door	17 fire	27 lamp	37 city	47 *BALLS*
8 alone	18 watch	28 *SCREW*	38 *QUEER*	48 *PUSSY*
9 good	19 *SNATCH*	29 paper	39 water	49 *BOX*
10 ride	20 drink	30 *PRICK*	40 *PIECE*	50 *LAY*

[a]Derived from Lefcourt, Gronnerud, & McDonald, 1973.

choosing between the responses relevant to each of the two meanings of the stimulus words.

Second, Lefcourt *et al.* (1973) had videotaped the facial and bodily movements made by subjects as they responded to the word list. These recorded data allowed for the observation of visible changes which indicated that subjects were becoming aware of the "odd" nature of the stimulus list. These data were like the "lightbulb effect" used in cartoons to show that a cartoon character has "seen the light."

In Figure 7 it is apparent that internal subjects were the first to show the predicted delay in response time. Internals exhibited excessive delays (greater than the 80th percentile of response times given to all non-double entendre stimuli) between the second and third double entendres, on the average. On the other hand, the average external did not exhibit his first excessive delay until the fourth or fifth double entendre. As a further point of interest, it can be seen that external–field-dependent subjects, the hypothetically most external individuals, were the very last to exhibit an excessive delay. These results indicate that external–field-dependent subjects were the last to become aware of the response choice afforded by the double entendres, or required more instances to become as cognizant of the double meanings as the more internal subjects.

Most significant of the facial expressions recorded were those of smiles and laughter. As illustrated in Figure 8, internals smiled and laughed decidedly more often than did externals as the sexual content of the list increased. As the list progressed then, internal subjects noticed the dissonant elements in the word list more quickly than externals and were bemused at their discovery. Other evi-

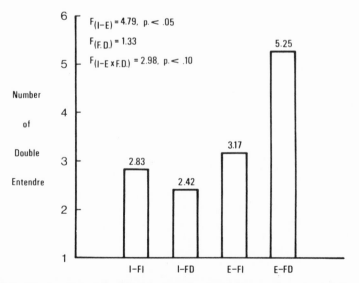

FIGURE 7   The number of first double entendre to elicit excessive response time (> 80th percentile). (Adapted from Lefcourt, Gronnerud, & Mcdonald, 1973.)

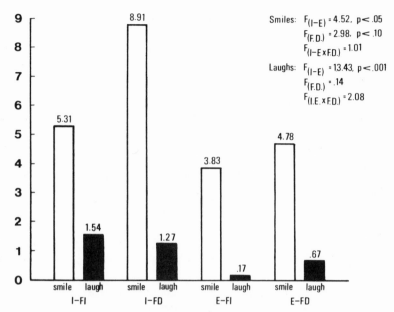

FIGURE 8 Frequency of smiles and laughs during the presentation of first sixteen words. Adapted from Lefcourt, Gronnerud, & Mcdonald, 1973.)

dence from this experiment included sex content of responses, startled looks, visible changes in attitude toward the experimenter, and so on. Generally, the results supported the contention that internals would be quicker at noting changes in the conditions about them and would be quicker to respond to their perceptions. In essence, internals were not as easily duped for as long a period as were externals due to a greater readiness to recognize and cognitively come to terms with change.

The second investigation has been recently reported by Wolk and DuCette (1974). Subjects were given the task of reading a story in which they were to search for typographical errors. After five minutes exposure to the reading material, subjects were given a test which asked for recall of names, dates, incidents, and other salient aspects of the story. The subjects, then, were assessed for the intentionally perceived (typographical errors) as well as for incidentally perceived material (story content). When this test was completed, the subjects were given the original story again and told to memorize the dates contained in it. This provided a second measure of intentional learning. After five minutes subjects were asked to recall the dates, and were also asked to remember the names contained in the story, a second measure of incidental learning.

Wolk and DuCette found that internals did consistently better than externals on both intentional and incidental learning measures. Internals found more typographical errors, recalled more story content, recalled more dates when instructed to, and recalled more names when not directed to do so than did externals. In each case the differences were statistically significant.

Wolk and DuCette replicated these results in a second experiment in which they varied several aspects of the task directions. The results from both experiments allowed the investigators to conclude that internals are more "perceptually sensitive":

> Because the intentional task required a quick, efficient scanning strategy, it is suggested that basic, pre-attentive processes differentiate the internal from the external. . . . The more interesting aspect of these studies, of course, was the fact that internals demonstrated higher levels of incidental learning. Incidental learning is a phenomenon dependent on the acquisition of less prominent aspects of a stimulus array, and since such acquisition has been interpreted as the product of a more attentive and organizing system, it follows that the internal differs from the external in the manner in which he organizes and uses information [Wolk & DuCette, 1974, p. 98].

Another point of interest was made by these writers. When scores on intentional and incidental learning were correlated within the internal and external samples, the differences between groups were marked. Intentional and incidental learning were strongly related for internals ($r = .80$ and $.58$) but were weakly related for externals ($r = .09$ and $.06$) in the first study. In the second experiment these findings were repeated with internals exhibiting high magnitude correlations between the two types of learning ($r = .52$ and $.58$), while externals exhibited the same low magnitude relationships as previously ($r = -.11$ and $.20$).

When these researchers clearly instructed subjects that they should be attentive to the story content as well as to search for typographical errors, externals showed a positive relationship between intentional and incidental learning ($r = .45$ and $.58$). Internals, on the other hand, showed no changes with the clarifying instructions. These authors concluded as follows:

> If the external can be characterized as an individual who, relative to the internal, possesses less active perceptual cognitive processes . . . but who also fails to use these processes efficiently, then extremely low correlations between intentional and incidental performance are meaningful. Thus, under a condition of ambiguity (Study I and the low cue explication condition of Study II), the external deals with the components of the task as if they were discrete. This lack of continuity in the task is a function of the external's failure to structure stimuli effectively, as well as an inefficient allocation of attention during the task itself. Because of this lack of continuity, how the external does on one task has no relationship to how he performs on the other. However, when the task is further explicated, that is, when attention is directed toward the significance of the content of the story, making the task more integrated, the relationship between intentional and incidental performance becomes extremely strong for the external and nonsignificantly different from that of the internal. Thus, it appears that the external does not make full use of his attentional system until stimuli are made more salient or prominent. For the internal such an explication is redundant, since his strategy has been to deal with the task in a more organized fashion all along. It is suggested then, that the external, relative to the internal, possesses a less active perceptual-attentive system (there are still significant overall performance differences in incidental learning between

internal and external Ss under the high-cue-explication conditions), but that the external also fails to use this system as efficiently as possible, especially under conditions of ambiguity [Wolk & DuCette, 1974, p. 99].

CONCLUSIONS

Our initial concern was to examine the processes that might allow an individual to resist pressures which could diminish his sense of responsibility. It was noted that internal control expectancies were associated with resistance to coercion, and an assertion was made that cognitive differences between internals and externals might account for differential responses to such pressures.

It would seem that the assumed differences in cognitive activity between internals and externals have been demonstrated. Internals have been found to be more perceptive to and ready to learn about their surroundings. They are more inquisitive, curious, and efficient processors of information than are externals.

Granted these differences, it may be asked whether an Eichmann or a Calley might have been more able to resist complying with horrendous requests had they been more able to think about the demands made upon them. Would Eichmann have so easily succumbed to his "Pontius Pilate" feeling had he been able to juxtapose his initial moral revulsion with the enthusiasm spurred by competition with his fellow bureaucrats?

Our data suggest that a self-acknowledged pawn such as Eichmann lacks the very cognitive processes that would enable him to examine and evaluate his choices and decisions. Unable to scrutinize his own responses and decisions, he may fail to even see that he has choices available to him. Lacking the perception of choice, he will yield easily to external pressures, be they for good or for evil.

6

Locus of Control and Achievement-Related Behavior

INTRODUCTION

In the preceding chapter it was concluded that certain cognitive activities were more characteristic of individuals with internal rather than external control orientations. One possible interpretation derivable from these findings is that locus of control is an epiphenomenon, a mere diagnostic indicator of a person's natural capacities for achievement; that is, the more intelligent and achieving a person is, the more likely he will perceive himself to be an active, effective person. The research described in this chapter fails to support a simplistic, one-to-one relationship between locus of control and achievement. As in most instances when a topic is closely scrutinized, the observed relationships are found to be anything but simple and conclusive. Nevertheless, it will be apparent that locus of control plays a mediating role in determining whether persons become involved in the pursuit of achievement.

The link between locus of control and cognitive activity appeals to common sense. In like fashion, common sense suggests that a disbelief in the contingency between one's efforts and outcomes should preclude achievement striving. Without an expectation of internal control, persistence despite imminent failure, the postponement of immediate pleasures, and the organizing of one's time and efforts would be unlikely. Common sense would dictate that these characteristics, essential to any prolonged achievement effort, will occur only among individuals who believe that they can, through their own efforts, accomplish desired goals; that is, individuals must entertain some hope that their efforts can be effective before one can expect them to make the sacrifices that are prerequisite for achievement.

The most commonly observed achievement activity occurs within school settings. Scholastic achievement requires that a youngster persist at activities such as reading when his immediate inclinations might be to play, daydream, or to socialize with his friends. The promise of future satisfactions to be derived from achievement effort is not always obvious. At the same time, daily activities which often consist of repetitive practice are often not enjoyable and interfere with more pleasuresome activity. Schools then, like other situations that offer the opportunity for achievement, require a degree of self-management, conscious effort, and the sacrifice of immediate pleasures for the possibility of future goal attainments. As noted above, such sacrifice is improbable if a person entertains doubts about his own potential effectiveness.

Until recently, failure in scholastic achievement was most commonly attributed to a low level of intelligence and success to a high level. In the 1950s, however, an interest developed in the motivation for achievement spurred on by the research of David McClelland and his colleagues (McClelland, Atkinson, Clark, & Lowell, 1953). The impact of this research was to alert psychologists to the importance of variables other than those assessed by intelligence tests for the prediction of achievement activity. In addition, it helped to define the components of achievement activity itself so that achievement became less singly identified with course grades.

It was not until the 1960s, however, that personality characteristics relevant to scholastic success began to receive extensive attention. With the rise of "Black Power" and other sundry "power groups," the public began to become aware that various ethnic, racial, and cultural groups differed considerably in their perceptions of many social institutions.

For some, schools had always seemed to be a channel to opportunity, and the sacrifices required for schooling were deemed worthwhile. For others, schooling seemed to lead nowhere and served little function other than that of a detention center. For the latter, the self-control and discipline demanded in school seemed oppressive and self-destructive. Among persons holding such views, achievement within schools would be unlikely. In fact, slum-dwelling blacks who find the schools to be remote, repressive, middle-class institutions, usually score at least one standard deviation below white averages on tests of intelligence and achievement.

The Coleman report (Coleman, Campbell, Hobson, McPartland, Mood, Weinfeld, & York, 1966) directed further public attention to the personality characteristics among the disadvantaged which limit their potential for achievement. Of immediate importance were the findings concerned with expectancies for control. Coleman and his colleagues found among nonwhite children that achievement was best predicted by a measure of the child's belief that academic outcomes were determinable by his own efforts. Pettigrew (1967) summarized the relevant findings from the Coleman report as follows:

Student attitude variables are surprisingly strong independent correlates of performance for all groups of children, though different attitude measures predict white and Negro achievement. An "academic self-concept" variable-measured by such items as "How

bright do you think you are in comparison with the other students in your grade?"—proves more significant for white performance. But a measure of "fate control" or "control of the environment"—indicated, for example, by disagreeing that "good luck is more important than hard work for success"—is much more important for Negro performance. . . .

Having experienced an unresponsive environment, the virtues of hard work, or diligent and extended effort toward achievement, appear to such a (minority) child unlikely to be rewarding. As a consequence, he is likely to merely "adjust" to his environment, finding satisfaction in passive pursuits. It may well be, then, that one of the keys toward success for minorities which have experienced disadvantage and a particularly unresponsive environment—either in the home or the larger society—is a change in this conception [Coleman, Campbell, Hobson, McPartland, Mood, Weinfeld, & York, 1966, p. 321; cited in Pettigrew, 1967, pp. 283–284].

In a later volume, Coleman (1971) himself reflected upon his findings with the sense of control:

The importance of attitudes such as this is the effect such an orientation toward the environment can have on other resources, by creating an active, driving stance toward the environment rather than a passive one. Suggestive evidence of its importance is provided by a striking result: those 9th-grade Negroes who gave the "hard work" response scored higher on the verbal achievement test, both in the North and South, than those whites who gave the "good luck" response, even though . . . the average Negro scored from 2.7 to 3.8 years (in different regions) behind northern urban whites and nearly as far behind whites from other regions. Although it is not possible to separate cause and effect by use of these data, this result does suggest an enormous potential impact of such an orientation in the creation of other resources [Coleman, 1971, p. 28].

These data and theoretical discussions are not without precedent; the problems had previously been diagnosed in similar terms and then ignored. Arnold Rose (1948), in a condensed version of Gunnar Myrdal's *The American Dilemma* (1944), wrote that

The ambition of the Negro Youth is cramped not only by the severe restrictions placed in his way by segregation and discrimination but also by the low expectation from both White and Negro society. He is not expected to make good in the same way as the white youth. And if he is not extraordinary he will not expect it of himself and will not really put his shoulder to the wheel [Rose, 1948, p. 218].

In a later section of his book, Rose had commented upon the sense of futility held by blacks in regard to their striving for valued goals, and the resultant antisocial or lethargic stance with which they met life's challenges:

Negroes know that all the striving they may do cannot carry them very high anyway. They often feel "Oh, you might as well make the most of it; what the hell difference does it make?" In this spirit, life becomes cheap and crime not so bad. Thus both the lack of a strong cultural tradition and the caste-fostered trait of cynical bitterness combine to make the Negro less inhibited in a way which may be dangerous to his fellows. They also make him more lazy, less punctual, less careful, and generally less efficient as a functioning member of society [Rose, 1948, p. 302].

These writers and others (see Chapter 2) have intimated that the sense of personal control is an important determinant of achievement-oriented behavior. In the subsequent section of this chapter, the empirical relationships between achievement activities and the locus of control will be explored through a review of relevant investigations.

ACADEMIC PERFORMANCE

The first investigation to implicate locus of control in the literature concerned with achievement-related behavior was reported by the late Vaughn Crandall and his colleagues at the Fel's Research Institute (Crandall, Katkovsky, & Preston, 1962). These investigators used a number of personality measures in the hope of predicting achievement behaviors as they were reflected in free-play activities, the Stanford–Binet Intelligence test, and the California Achievement Tests. The predictor measures included a TAT measure of need for achievement, a scale for assessing manifest anxiety, the children's own statements regarding their concern for intellectual attainment, expectations of intellectual success, intellectual achievement standards (the child's own definition of success), and the first form of the Intellectual Achievement Responsibility Questionnaire (IAR). The latter (see Appendix) was comprised of 36 items—half concerned with successes and half with failures in the realm of achievement. Children had to choose whether the success or failure experiences described in each item would be primarily the results of their own behaviors or would more likely be caused by external agents such as parents or teachers.

Two free-play measures of achievement were based on the children's activity during a week's day camp at the Fel's Institute. The main playroom was stocked with miscellaneous play materials and equipment conducive to achievement activity. Each child was free to play with whatever he chose. Examples of intellectual and achievement materials were books pertaining to scientific and other informational topics, games involving intellectual competition such as checkers, chess, chinese checkers, and so on, jigsaw puzzles, form boards, arithmetic and reading flash cards, and a variety of intellectual quiz games.

Observers time-sampled the children's playroom activity, recording both the amount of time spent in intellectual activities during 15 observation periods, and the intensity of striving in these activities. The latter was defined in terms of the degree of concentration that the child exhibited, ranging from "playing around" to strong concentration with the use of observable personal achievement standards.

Of all the measures employed by Crandall *et al.* (1962), the IAR proved to be the most strongly related to the time spent in intellectual free-play activities ($r = .70, p < .05$) and to the intensity of striving in those activities ($r = .66, p < .05$) among the boys. For the girls, on the other hand, the rs were .01 and .00,

Gender differences.

respectively. In other words, for boys the attribution of responsibility was of considerable importance for predicting achievement activity, while for girls it was totally irrelevant. Personal value for intellectual attainment was the only significant correlate of free-play achievement activity for girls.

When performance on intelligence and achievement tests was correlated with the predictor variables, a similar pattern of results was obtained. The IAR was significantly related to each test for the boys ($r = .52, p < .05$ with intelligence; $r = .51, p < .05$ with reading achievement; $r = .38, p < .05$ with arithmetic achievements), but totally unrelated for girls ($r = .00; r = -.03; r = -.13$, respectively). With regard to the achievement test results, these correlations are of signal interest in view of the fact that none of the other achievement-related variables (need—achievement, anxiety, value for intellectual achievement, etc.) were significantly related to achievement tests for the sample of boys.

These findings have been replicated, but with some differing results (Chance, 1965). With the same tests (Stanford—Binet, California Achievement Tests, and the IAR) Chance found a similar pattern of correlations among third-grade boys ($r = .34, p < .01$ for intelligence; $r = .50, p < .01$ for reading achievement; $r = .46, p < .01$ for arithmetic achievement; and $r = .56, p < .01$ for spelling). However, where Crandall *et al.* (1962) had found the IAR questionnaire to be unrelated to achievement among girls, Chance (1965) found $rs=.34, .45, .51$, and .38 between IAR and each of the respective achievement variables. These correlations were all significant at the $p < .01$ level except for the first, which was significant at the .05 level.

In a later study conducted by the Fel's group (Crandall, Katkovsky, & Crandall, 1965) further data were presented concerning the IAR—achievement behavior corelations. Among third-, fourth-, and fifth-grade boys and girls, the IAR was found to be significantly related to reading, math, language, and total achievement test scores from the Iowa Tests of Basic Skills. In addition, report card grades for this sample were associated with the IAR.

The IAR, as may be seen in the Appendix, contains two sets of questions. Half pertain to attributions of responsibility for success (I^+) and half to attributions for failure (I^-). The total IAR score noted above is a composite of internal attributions for both success and failure. When Crandall *et al.* (1965) evaluated I^+ and I^- separately with the achievement measures, the results were a bit perplexing and inconsistent. All achievement test measures and report card grades were highly related to I^+ (rs ranging between .40 and .60) "indicating that the greater the young girl's" (grades 3 and 4) "sense of responsibility for her academic success, the more successful she is likely to be [Crandall, Katkovsky, & Crandall, 1965, p. 107]." Scores for attributions of responsibility for failure, on the other hand, were significantly correlated with the nine measures for fifth-grade boys with rs ranging from .34 to .53. To make matters more confusing, in grades 6, 8, 10, and 12 scores on the California Achievement Tests were only occasionally related to IAR scores, though significant relationships between report card grades and total IAR scores did obtain in these upper grades. A ninth-grade sample, on the other hand, produced results similar to the first study by Crandall, Katkovsky, and Preston (1962). Achievement test results for reading, language, and arithmetic were related to I^+ scores for boys (rs in the .50s) but not for girls.

Thus, what seemed to be a simple set of predicted relationships is obviously beset by unpredictable variability. Girls and boys differ in characteristics that were assumed to be related to achievement. Age or grade level makes a difference, and we are now introduced to the concept of locus of control as a less pervasive or general characteristic—one that may be usefully divided into components such as responsibility attribution for success versus failure experiences.

Other investigations have borne out the fact that achievement activity is associated with locus of control, though not consistently and not without occasional paradox. Franklin (1963) found internality related to the amount of time that high school students spend in doing homework; James (1965) reported that internals were more persistent in their attempts to solve complex logical puzzles, and the Fel's investigators (McGhee & Crandall, 1968) have found replication of their previous findings: report card grades of both boys and girls were associated with locus of control in samples of children in grades 6 through 12, and scores on the Iowa Achievement Test favored internal children in the third through fifth grades with some variability between sexes and the I^+ and I^- measures.

In a study using large samples of eighth and eleventh graders from the Chicago schools, Lessing (1969) found that the Sense of Personal Control, as assessed by Strodtbeck's Personal Control Scale (see Appendix), was correlated with grade point averages even when IQ scores had been statistically partialled out; that is, among both the more and less intelligent pupils, a sense of personal control was related to actual school achievement. Harrison (1968) likewise has found that a sense of personal control allowed for some prediction of success in school regardless of the socioeconomic status of the childrens' homes; that is, an internal orientation as measured by Harrison's own "View of the Environment Test" predicted academic success among both advantaged and disadvantaged children. This result as well as that reported by Lessing indicate the strength of the association between the sense of control and achievement behavior. Intelligence test performance and socioeconomic status have a reliable and robust relationship with achievement criteria. That a sense of control, measured by different devices, can add to the already high magnitude relationships between socioeconomic class, IQ, and achievement behavior attests to the value of locus of control in formulas devised to predict achievement behavior.

Nevertheless, as the studies noted above and others reveal (Katz, 1967; Nowicki & Roundtree, 1971; Stephens, 1972), the relationships between various measures of locus of control and achievement behaviors are often riddled with inconsistent and, as Stephens puts it, "weird" results.

DEFERRED GRATIFICATION

As in the case of achievement tasks, locus of control is often relevant to the willingness to defer gratification, though not always in a consistent and comprehensible way. It is plausible to assume that internals are more accustomed than are externals to engaging in the execution of long-range plans. The very process of planning and working for distant goals would only seem to be sufferable if the individual believed that he was able to determine the results of his efforts. Externals, by definition, find the paths from initiation to completion fraught with uncertainty. Why should externals postpone immediately available pleasures for distant goals when daily events occur by some nonpredictable design?

Distant goals require the sacrifice of immediate pleasures. As a writer struggles to express his ideas on paper, the warm sunshine of an Indian summer day can act as a seducer, pulling him away from a long and possibly fruitless enterprise. Throughout a morning's work he may find himself beleaguered with ambivalence—angered at his self-inflicted deprivation and yet pleased at being able to control himself during the continuation of long-range activity. On other days, the weather and scenery might not be as tempting and competitive. Throughout any long-range endeavors, however, such as research activity or book writing, temptations inevitably recur. To maintain the pursuit of distant goals despite temptations seems to require a faith in the reliability of one's perceived world. A disbelief in the predictability of daily events should hinder one in long-range activities and, most clearly, if a person were an unpredictable commodity to himself he would never embark on lengthy drawn out enterprises. In short, to keep himself pressing on to the completion of a long-range project, a person must feel relatively secure in his world. He can not feel as if uncertainty stalks his every move. He must know with some level of assuredness that he is reliable and capable of bringing off his project. Otherwise, the tempting sirens of daily pleasure would be all too compelling. If tomorrow brings us possible calamity, why deny one's self today's small offerings of pleasure. Long-range efforts become absurd if we know that we will be encountering daily uncertainty.

Empirically, the first support for this assumed connection between locus of control and the ability to defer gratification was reported by Bialer (1961). This investigator offered a series of choices to a sample of children. The first involved a choice among improbable wishes—"Which would you rather have, the best kind of automobile right now—and you'd have a driver's license and know how to drive it too—or have the automobile and a million dollars a year from now?" The choice of the "automobile now" resulted in a score of 0, the delayed choice, a score of 1. The second choice was between a single piece of candy now, or four pieces on the following day. The former choice resulted in a 0 score, the latter, a score of 2. Third, the subjects were offered a choice between a penny now, and ten pennies on the following day. If the child opted for the single penny he received a 0. If the child selected the 10 pennies delayed choice, he was then given a further option of 5 pennies now or 10 tomorrow. If the child initially chose the 10 pennies versus the one, he received an additional 2 points. If he subsequently elected to take the 5 pennies he was given no further credit. If, on the other hand, he chose to delay again, selecting the 10 pennies, he received an additional 2 points. In total then, the child could receive scores between 0 and 7.

With his own locus of control scale (see Appendix) Bialer found that deferred gratification was associated with internal locus of control ($r = .47, p < .001, N = 89$). As Bialer had suggested when offering his hypotheses, internals seemed to be better able to maintain the tension associated with delays than externals. "The more mature child, aware that his own efforts can forestall failure, and being able to maintain the tension generated by the postponement of immediate need satisfaction, should therefore choose to defer his gratification [Bialer, 1961, p. 306]."

While this investigation offers some support regarding locus of control and the ability to defer gratification, it differs sufficiently from the previous discussion to warrant attention. In the Bialer study, the delay was just between "now" and "tomorrow." No effort or persistence at a difficult and frustrating task was required. The choices presented to the children were certain and were not contingent upon specific efforts. On the other hand, Bialer's procedures and those circumstances confronting a person engaged in long-range pursuits offer similar challenge to one's ability to overcome the tensions experienced in rejecting seductive diversions. The immediate small pleasure is tempting, a hurdle over which one must pass if he is to achieve larger goals. To ignore an immediate prize for more enticing prizes later involves the ability to resist temptations and to not feel excessive arousal at being denied immediate pleasures. It is this resistance to temptation that is common to the simple task of tolerating a delay of pleasure and the more complex persistence in the service of long-range goals.

The commonly employed technique of offering youngsters either a small prize immediately or a larger gift to be delivered at a later date was developed by Alvin Mahrer (1956) and used in an extensive set of investigations by Walter Mischel (1966). While much of the research with delay of reinforcements has involved simple waiting, as in Bialer's study, some of Mischel's more recent research has examined delayed reinforcement choices when effort and work are required.

Mischel, Zeiss, and Zeiss (1974) presented findings from a series of investigations in which they made rather precise predictions of persistence with their own Stanford Preschool Internal–External Scale (SPIES—see Appendix). The SPIES, like the IAR, is composed of two measures, the perception of control over positive and of negative outcomes. In this study, Mischel et al. (1974) predicted that internality for success experiences would be associated with persistence at activities that are instrumental to the achievement of valued goals, while internality for failure would be associated with behavior aimed at avoiding aversive circumstances.

In each of the studies described by Mischel *et al.* (1974) children were offered the option of accepting smaller rewards early in the task or to work for some period of time for larger prizes. The subject could quit at any time and accept a less valuable prize. While the results in several studies varied in magnitude, Mischel *et al.* (1974) found that internality for success was positively related to persistence in the effort to obtain the larger, delayed prizes. Correlations ranged from .32, $p < .05$ to .66, $p < .01$ in different experiments with varying conditions.

In contrast, internality for failure was not related to the children's choices. However, when the children anticipated losing previously earned prizes depending upon their performance on selected tasks, internality for failure proved to be more valuable than internality for success in predicting practicing behavior; that is, children who perceived themselves as responsible for failure experiences practiced more than children who felt external with regard to failure on a task in which they anticipated punishment for failure. These findings occurred, however, only when the children were led to believe that they could avoid punishment by their performance. On the other hand, when performance and outcome (whether or not the child would lose his prizes) were made to seem to be unrelated, the correlations were significant but negative—that is, the more internal for failure, the less likely were children to practice. Thus, Mischel and his colleagues were able to demonstrate the value of considering causal attributions of success and failure independently—and of differentiating between work aimed at securing delayed rewards and work aimed at averting delayed punishments.

Other researchers have reported interesting if mixed findings with regard to locus of control and deferred gratification. Zytkoskee, Strickland, and Watson (1971) for instance, found that both an internal locus of control and self-imposed delay of gratification were common to certain groups. Poor Southern black children were decidedly more external than poor white children on Bialer's locus of control scale and were more likely to choose immediate smaller reinforcements than to wait three weeks for larger reinforcements. However, despite the coincidence of locus of control and a preference for delayed larger reinforcements, the two variables were unrelated for the total sample ($r = .09$, n.s.). In a subsequent study, Strickland (1972) contrasted the results of black and white children as they chose between immediate and delayed reinforcements that were offered by white or black experimenters. Three hundred sixth graders completed the locus of control scale and were then offered one 45-rpm record immediately or three records if they could wait for three weeks. Strickland found again that blacks were more likely to choose the immediate, smaller reinforcement and were more external than whites as assessed by the Nowicki–Strickland locus of control measure. However, Strickland found that the choice of immediate reinforcement was most prevalent among blacks when the experimenter was white. With a black experimenter delayed reinforcement preference increased considerably. For the black children the race of the experimenter was more important than locus of control for predicting their willingness to delay. However, for white children, locus of control was significantly related to choice behavior. Those white children who chose the delayed were more internal than were those who preferred the immediate reward. In further contrast to black children, the experimenter's race was irrelevant to white childrens' choices.

Strickland's research indicates that the willingness to defer gratification is predictable, though often on the basis of rather different variables. If security or predictability of one's phenomenological life is of the essence for choosing the delayed reinforcement option, as suggested earlier, then the differences between black and white children are worthy of comment. For the white child, the major source of unpredictability may be thought of as personal; that is, he is sufficiently differentiated from others that his experiences are viewed as personal—what happens to him is, relatively speaking, a function of his unique identity. Thus, when offered a choice of immediate versus delayed reinforcement, his self-perceptions predict his responses. For black children, however, *race* is a major source of uncertainty. When interacting with white persons, variation in responses is great enough that prediction of the white person's behavior becomes problematic. An optimal strategy to adopt when predictions approach a 50–50 level of accuracy is to stick with one bet. Then one will be correct at least half the time. In this case, a black child might conclude that he should never trust a white person to follow up on his word. If a 50–50 choice level may be considered generous in describing the actual probability with which whites fulfill promises to blacks in the South, it is possible to imagine how black children can develop a pervasive cynicism regarding promises from whites. When offered a choice of immediate versus delayed reinforcement then, the black child looks outward—the source of unpredictability and insecurity is to be read in the social milieu, and not within one's self. If the source of the promise is a white person, one set of expectancies is evoked; if a black person, then it is another. In either case, the elements in the external world bear considerably more importance for determining stability than would the personal characteristics of the black child.

Strickland's data then, contain some support for the predicted locus of control-deferred gratification relationship. If our conjectures should prove accurate with further empirical tests, the limits of this relationship may also be evident.

Other investigators have reported varying results as they have compared locus of control and deferred gratification. Walls and Smith (1970) found locus of control related to delay among second- and third-grade children, with internals choosing to wait for a 7¢ as opposed to an immediate 5¢ prize. However, Walls and Miller (1970) found no relationship between the same variables with a small sample of welfare and vocational rehabilitation clients though both variables were related to education. The more educated the individual the more internal they were and the more likely they were to prefer delayed reinforcement.

In the earlier study by Walls, a relationship between locus of control and accurate judgment of time was also noted. Internals were more accurate in judging the lapse of a minute. This finding is of interest in view of the fact that correctness of time judgments is, in turn, related to the preference of delayed reinforcement (Mischel, 1961).

In addition to time judgment, locus of control has been found to be correlated with other time-related measures such as future time perspective (Platt & Eisenman, 1968; Shybut, 1968). Internals are found to have a longer future time perspective than externals; in addition, measures of "delaying capacity" (Kagan's Matching Familiar Figures Test and the Porteus Mazes) have been found to be related to locus of control (Shipe, 1971), though the device of immediate versus delayed reinforcement was not related to either the delaying capacity or locus of control measures in Shipe's investigation.

A more convincing test of the linkage between I—E and deferred gratification has been reported by Erikson and Roberts (1971). Institutionalized adolescent delinquents were offered the opportunity of attending public school away from the grounds of a reformatory. While this is often an exciting diversion for reformatory inmates, this opportunity was granted with the proviso that the inmates' release from the institution would therefore be delayed. Subsequently those who decided to accept this "real" delay choice were compared with those who did not. One question served as the dependent measure of locus of control. The inmates were asked "Why is a boy transferred to 'ZB' (the name of the delayer's cottage)?" The responses were then classified as internal, external, or neutral. Only one internal attribution was made among nondelayers (or 5%) in comparison to 8 (40%) from among delayers. In this very real delay choice, then, internal attributions were more common among those selecting the delayed choice. While this measure of locus of control consists of only a single answer, the veridicality of the choices lends credence to the obtained findings.

Overall, the studies reported support the hypothesized relationship between I—E and the ability to defer gratification in the pursuit of long-range goals. As with achievement, the findings are often variable and at times inconsistent. Various measures such as time judgments and time perspective are linked with locus of control, and it is evident that the perception of time is of some importance in man's perception of himself as an active determiner of his life's directions. A most apt reference, then, may be made to a study by Melges and Weisz (1971) of suicide.

These investigators enlisted the aid of 15 patients who had recently made a serious attempt at suicide. The subjects were asked to reconstruct, as vividly as possible, their state of mind immediately prior to their suicide attempt. Subjects were left alone with a tape recorder into which they delivered a soliloquy—imagining aloud the events and feelings that were going through their minds at that time. Both before and after the soliloquy, subjects were asked to complete a number of tests, including one concerned with the perception of one's future, and another, a modified version of the I–E scale.

Melges and Weisz were interested in examining the effects of increased suicide ideation on each of the measures administered. Their findings indicated that increases in suicidal ideation were associated with changes toward a more negative evaluation of the personal future, changes toward less internal control, and changes toward less extension of a span of awareness into the future. In turn, changes in personal future were correlated in the same direction with changes in the locus of control (r = .61, $p < .05$)—a negative outlook for the future and external control expectancies were associated with each other and, in turn, with suicidal ideation. Likewise, changes in locus of control correlated with changes in temporal extension into the future (r = .63, $p < .05$) suggesting that as one feels increasingly helpless, one's span of awareness of the future diminishes.

The image of an individual at the brink of suicide, being trapped within the immediate moment, for whom the future has ceased to exist as a meaningful and positive force, and who experiences himself as helpless to effect his fate toward positive ends spells out for us the potential mutual relevance of the perceptions of time and personal control, as have few of the previous studies.

CONCLUSIONS

Research has indicated that achievement effort and the willingness or ability to tolerate delays in the attainment of reinforcements are related to the perception of causation. Confirmatory evidence exists in support of the hypotheses and observations of writers who have attempted to explain the high rates of failure at achievement activities among members of particular groups. Individuals who develop with little expectation that life's satisfactions and misfortunes can be determined by personal efforts have been less apt to exert themselves or to persist over lengthy time intervals in the pursuit of distant goals; and, as if oftentimes has been contended, such exertion and persistence are the sine qua non of achievement activity. Congruent with the more personal material discussed in earlier chapters, research findings indicate that the engagement in achievement activity or long-range skill-demanding tasks is unlikely if one views himself as being at the mercy of capricious external forces. However, as noted above, the empirical data are not often without paradoxical inconsistencies or failures at replication.

Recently, a number of investigators have explored causal attributions in a manner which may help to improve upon previous attempts to predict achievement-related behavior. Most notable among these investigators is Bernard Weiner (Weiner, Heckhausen, Meyer, & Cook, 1972), who has with some success added

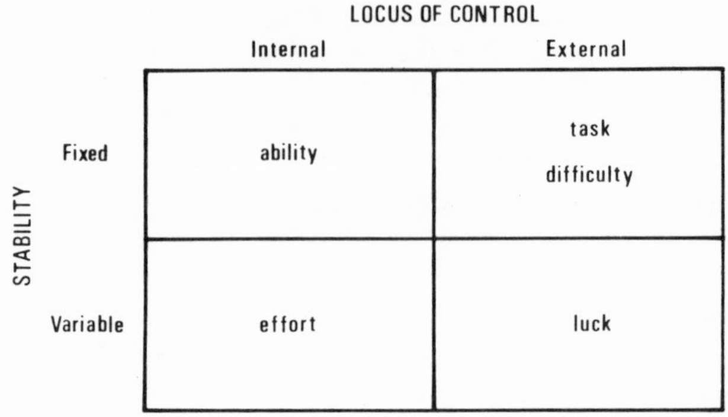

LOCUS OF CONTROL

FIGURE 9   The perceived determinants of success and failure. (From Weiner, Heckhausen, Meyer, & Cook, 1972.)

a dimension of "stability" which interacts with locus of control in predicting achievement behaviors. Figure 9 contains a diagrammatic presentation of Weiner's use of locus of control and stability of cause. As can be seen in this figure, internal causes are divided into those that will remain fairly constant, such as ability, and those that can vary for different reasons, such as effort. External factors may also be variable, such as luck, or stable, such as in the difficulty of a given task.

On the basis of experimentation with this paradigm, Weiner has tentatively concluded that the choice to engage in achievement activity is mediated by internal variable factors, such as effort, which generate positive feelings; that is, persons who perceive that outcomes in achievement activities are determined by variations of their own effort, as opposed to constant ability, will find more pleasure engaging in their pursuits. Second, persistence despite failure is said to be more likely if the causes of failure are seen as variable. If bad luck or lack of effort have been responsible for failure, then hope for improvement is still plausible. If one's effort or luck has been failing then change is possible and perhaps imminent.

It is premature to speculate at this point on all the yet to be reported data that such a discriminating use of perceived control measures may offer. Let it suffice to say that the uncertainties generated by inconsistent findings noted in this chapter may be resolvable as the more differentiated conceptions of locus of control offered by Weiner and Mischel are employed in the attempt to predict achievement-related activities.

7

Fatalism and Psychopathology

INTRODUCTION

When theoreticians attempt to discuss phenomena associated with psychopathology, readers are often left with a strong feeling of disappointment. Words in general fail to convey the reality of terror experienced by a person undergoing an acute psychotic reaction, and the abstruse terminology of systematic theories seems far removed from such immediate experience. Nevertheless, in this chapter, we will try to discuss abnormal phenomena using social learning terminology with the hope that the utility of our terms will compensate for their lack of dramatic appeal.

In previous chapters, several attempts to link locus of control with various abnormal phenomena were described. Conversion reaction hysteria, ulcers, and attempted suicide were each discussed as possible concomitants of an external locus of control. The temptation to link locus of control to a wide range of clinical phenomena derives less from a wish to reduce such variety to simplicity than it does from a desire to respond to the compelling symptoms presented by psychiatric patients.

Paranoid individuals speak of their omnipotence, if given to grandiosity, or of abject helplessness before powerful external forces if suffering with delusions of persecution. Manic patients openly dismiss suggestions that they are subject to limitations as are other human beings; depressives insist on their helplessness through inaction if not words; while antisocially inclined individuals may explain their deviations as necessary and beyond personal choice. Despite these differences among clinical groups as diverse as depressives and delinquents, there is often commonality in the tendency to eschew personal responsibility and to espouse an external control orientation.

An immediate connection between psychopathology and locus of control can be established if we return to the topic of deferred gratification discussed with regard to achievement in the preceding chapter. Freud and Mowrer each developed constructs pertaining to the ability to defer gratification—impulse control, time binding, and temporal integration which, in turn, have been used to explain certain pathological phenomena. Mowrer (1950) has explained the persistence of self-punitive behavior common among persons suffering with neuroses in terms of a failure of temporal integration. Neurotic behavior, characterized by expedient actions designed to avoid immediate challenges at the cost of later suffering, is said to occur among persons who are ignorant about the contingencies between immediate behaviors and later consequences. Mowrer (1950) described his position in the following manner:

> ... if an immediate consequence is slightly rewarding, it may outweigh a greater but more remote punishing consequence. And equally, if an immediate consequence is slightly punishing, it may outweigh a greater but more remote rewarding consequence. Living organisms which are not skilled in the use of symbols are severely limited in their capacity to resolve behavioral dilemmas of this kind and may, as a result, continue indefinitely to manifest so-called nonintegrative behavior. But by introducing the time element ... it is possible for us to escape from the dilemma which such behavior presents from a theoretical standpoint.
>
> The prodigious capacity found in normal adult human beings for using symbols, i.e., for "reasoning," seems to have what is perhaps its greatest utility in enabling the individual to bring the remote as well as immediate consequences of a contemplated action into the psychological present and thereby compare and balance the probably (anticipated) rewards and punishments in a manner which enormously increases the chances that resulting behavior will be integrative. Such behavior is properly termed rational, in contradistinction to the prerational behavior seen in lower animals [pp. 453–454].

Deviations other than the neuroses have also been interpreted in terms of temporal integration. The psychopathic person, for instance, is said to be unable to delay gratifications:

> He cannot surrender a smaller present pleasure for a greater distant goal. He cannot abide routine for long and takes no comfort in a stable, recurrent regime of daily living. He has no sense of time, especially of the future; he senses only the here and now. This makes him prey to almost any momentary temptation or escape from tension, regardless of subsequent consequences. He simply cannot bridge the gap between present pleasure and future punishments. In this respect he is similar to the neurotic who buys a temporary respite from anxiety now at the cost of greater punishment later [Buss, 1966, p. 435].

The schizophrenic, too, is said to be lacking in a sense of time. When asked to estimate the passing of a second, for instance, schizophrenics commonly make gross overestimates (Maher, 1966). Correctness of time judgments has, in turn, been found to be related to the ability to defer gratification (Mischel, 1961).

Different forms of personality malfunctioning, then, are associated with the inability to defer gratification; and as previously indicated, this inability is more common among persons who hold external control expectancies than among those who perceive themselves to be active determiners of their own fate. Therefore an indirect link, at least, may be assumed between locus of control and the incidence of psychopathology.

Other research reviewed in previous chapters offers support for this hypothesized association. That internals are more cognitively efficient, more alert to the potential meanings of their experiences, and less easily coerced by environmental forces attests to the importance that an internal locus of control has for effective coping behavior. If a person is able to quickly assess the options available to him in a challenging situation, he should be able to cope more effectively than if he were less astute about his choices; and, if he believed that he was able to effectively act in his own behalf, even aversive consequences would have a less debilitating effect upon him.

What is being suggested in this introduction, then, is that locus of control is pertinent to the development of personality malfunctioning if only in an indirect manner. That one tries to cope with challenges to his sense of well being, or simply persists in the belief that he can cope with adversities bodes well for an individual. Avoidant, defensive mannerisms associated with deviance and malfunctioning should be less evident among those persons who continue to strive after valued goals in the belief that their efforts are meaningful and effective.

DEFENSIVENESS

On the basis of our previous discussion it would seem unlikely that internals are more defensive or avoidant of challenges than externals. Nevertheless, the results from a number of studies have led certain investigators to conclude exactly that, and the reasoning used to draw this conclusion does not lack plausibility.

If one commonly attributes cause for outcomes to one's personal characteristics, then outcomes are self-relevant; that is, one's successes and failures are meaningful for learning about one's self. If one customarily attributes causality to external events, successes and failures should be of little matter as far as one's self-regard is concerned. Failures can be "explained away" in terms of "others," "circumstances," or fate. For the external, outcomes should be of little relevance to self-evaluation since one's self is not held responsible for those outcomes. Consequently, it is reasonable to assume that externals will have little need to defend themselves against failure, but that internals may resort to various subterfuges if they are to retain self-respect subsequent to failure experiences. Empirically, this position received some support in a dissertation by Efran (1963) in which internal high school students were more likely to have forgotten their failures than were external students.

A second study (Lipp, Kolstoe, James, & Randall, 1968) likewise suggested that internals might be more defensive than externals. Physically handicapped internals and externals were compared in terms of perception thresholds for stimuli relevant to physical handicaps. Subjects were shown micromomentary exposures of a series of slides, some of which portrayed physically handicapped persons. These investigators predicted that internals would have lower thresholds (would recognize handicap-relevant stimuli more quickly) than externals. However, their data offered anything but a straightforward confirmation of this hypothesis. With subjects divided into three groups on the basis of James' I—E scale, handicapped externals were found to be the quickest, handicapped internals second, and a sample of midrange I—E-handicapped subjects were the slowest to recognize the stimuli in pictures of disabled persons.

These data do not lend themselves to easy interpretation. If nonrecognition of disabled figures by disabled persons is construed as an instance of defensiveness, then externals can be said to have been the least, though internals were not the most defensive. However, if we do not equate defensiveness with slow recognition of crippled figures but rather we speak of "selective attention to relevant stimuli," these findings can be reinterpreted in a very different fashion. If a handicapped person comes to terms with his disability, managing through compensatory mechanisms to circumvent many of the difficulties resulting from the particular handicap, the sense of being handicapped may diminish in salience; that is, when individuals suffer an infliction their attention may readily focus upon the resulting disability for some period of time such that fellow sufferers may be sought for comparison and commiseration. However, as one comes to terms with his disability, the salience of the handicap should decrease, and previous interests and pursuits of the individual should reemerge. The amount of time expended upon thinking and speaking about the disability should then become negligible in contrast to the active interests in one's life. In fact, the term hypochondriasis is often used as a term of opprobrium to depict the state when a reemergence of nonhealth-related interests fails to occur. Thus, the handicapped individual who has learned to cope with his disability may become less attentive to handicap-related stimuli in himself or in others than he had formerly been.

In view of this reformulation, we may reinterpret the findings of Lipp et al. (1968) in terms of relevance instead of defensiveness. It would seem that for the external, the handicap still exists as a salient factor in his life. Disability-related stimuli are consequently perceived more quickly by externals perhaps because they have not been as able as internals to cope with the resulting impediments and, consequently, are still overly concerned with their disability.

Evidence from a survey conducted among a large group of nonhandicapped college students (MacDonald & Hall, 1971) lends some credence to the above interpretation. Students were questioned as to how debilitating various physical impairments would be with regard to their social relationships and feelings about themselves. Externals were found to anticipate more severe effects from physical disability than internals. The less pessimistic stance among internals may reflect the belief in their ability to cope with adversity. Armed with such a belief internals can then be expected to try to cope more adaptively with impairments than externals if mishaps were to befall them. Likewise, internals should not dwell upon their handicaps for as long a time as externals. If these assumptions are correct, then internals should not be as perceptually vigilant for handicap-related stimuli as externals. Be that as it may, however, the conclusions to be derived from the study by Lipp *et al.* (1968) are that externals are not more avoidant than internals of the handicap-related material, as had been predicted.

The first investigation designed specifically to evaluate defensiveness as a correlate of locus of control was reported by Phares, Ritchie, and Davis (1968). These investigators provided their subjects with interpretations from previously administered personality tests. The feedback was prearranged, however, each subject receiving the same eight positive interpretations, for example, "she is warm and is able to maintain very positive interpersonal relationships," and eleven negative interpretations, for example, "at times, sexual thoughts become a problem and make her doubt her maturity." While internals and externals did not differ in the discomfort that they reported when receiving the interpretations, externals were found to recall more of the 19 interpretations at the end of the experiment than were internals. Phares *et al.* (1968) had predicted that externals would particularly recall more negative interpretations than internals, since the externalization tendencies of the former should have eliminated possible anxiety responses to the negative interpretations. The results did not actually support this contention, however; externals were found to recall more interpretations, *in general, both positive and negative,* than were internals.

Though the data failed to clearly support the hypothesized sensitivity to failure among internals, Phares, in a subsequent study (Phares, Wilson, & Klyver, 1971), expressed the belief that the previous experiment had demonstrated that "external subjects recalled more of the negative information than internal subjects. Again it appeared almost as if an external orientation provided ready access to a defense and thus relatively less need to invoke forgetting as a response [Phares *et al.,* 1971, p. 285]." Though Phares and his colleagues continued to interpret their findings as if they illustrated a greater "need to forget" among internals, the results from the second experiment made this interpretation of defensiveness a bit less tenable. Internals and externals were failed on two tests which had been described as measures of intellectual functioning. The failure experiences were contrived by including insoluble puzzles among a legitimate set of puzzles. Subsequent to their failure experiences, each subject completed a scale of blame attribution. Under "fair" conditions, when the experimenters had remained quiet, allowing subjects to concentrate on the tasks, internals were decidedly less prone to blame the experimenter or testing conditions than were externals. However, when the experimenter engaged in distracting chatter throughout the test period this difference in blame attribution between internals and externals diminished.

As Figure 10 makes evident, externals were more likely to blame "circumstances" even when there was little justification for doing so. Internals blamed the experimenter for their failure when it was reasonable to do so but were less likely to hold him responsible when there was little justification for such blame.

These findings suggest that internals are more flexible than externals in their assignment of cause for failure. On the other hand, externals seem less able to accept the implications of failure, blaming external circumstances even when justification for doing so was negligible. This ability to remain flexible in the way one interprets failure experiences would not seem a likely characteristic of a person who was uncomfortable enough to "need to forget" those experiences. The inflexible approach of externals, however, does suggest the strain with which they assimilate their failures. While this study did not contain the opportunity to forget failure-relevant material, the overdetermined behavior commonly associated with repression as a defense mechanism was clearly less evident among internals than it was among externals.

Two other studies (Phares, 1971; Davis & Davis, 1972) lend additional evidence to that noted above. Externals seem to behave reliably in their readiness to eschew responsibility for failure while internals seem to be less driven to deny or avoid the information availed by failure experiences. In the Phares (1971) study subjects were administered four tasks described as measures of intelligence. Prior to performing on these tasks subjects were asked to rank the tests in terms of how much they wished to succeed on them. By arbitrarily limiting the

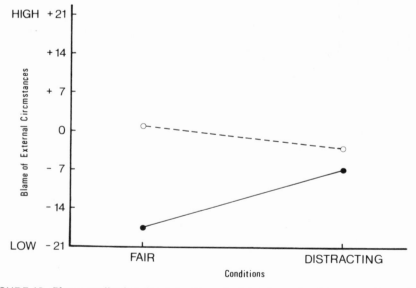

FIGURE 10   Blame attribution for internals and externals under fair versus distracting conditions. (●) Internal; (○) external. (From Phares, Wilson, & Klyver, 1971.)

time allowed for task completion subjects were failed on those tests ranked as either the first or second most preferred. Afterward, subjects were asked to rank the four tests again, this time with the benefit of firsthand experience. Phares found that internals did not shift in their evaluation of the tasks when given the opportunity to rerank their preferences. Externals, however, did shift their preferences away from those on which they had "failed." In two successive studies Davis and Davis (1972) found that internals readily blamed themselves for their failures more than did externals. Whether failure was incurred on an "intelligence" test comprised of a series of anagrams, or on a test of "social sensitivity" in which subjects had to predict another person's attitudes, externals were less likely to attribute cause for failure to themselves than were internals.

In these experiments, then, externals were found to be more likely to shift about in their judgments and evaluations, and more ready to denigrate tasks or to blame other persons for their failures than were internals. Such findings indicate, at least, that failure is more of an irritant to externals, causing them to reorder their constructions of events. The stability of internals, on the other hand, suggests perhaps that there is not the finality in failure that there is for externals. If a person maintains hope or a "wait until next time" approach, failure should not rest as heavy upon him. If he is responsible for his failures, then a bit more effort or familiarity with given tasks may enable him to exercise greater ability in subsequent trials with a task.

The data within these experiments concerned with response to failure do not provide us with firm interpretations. However, the ready acceptance of blame for failure by internals provides eloquent refutation of an argument positing a "need" to forget with all of the connotations of anxiety and defensiveness.

SELF-REPORTS OF PSYCHOPATHOLOGY

Most of the above discussion derives from research conducted in academic settings with contrived challenges for the persons involved. Whatever is lost in the artificiality of this research, however, is gained in the relevance of the content area of that research for such clinical problems as depression. Beck (1967), for instance, has recently written:

> The depressed patient is peculiarly sensitive to any impediments to his goal-directed activity. An obstacle is regarded as an impossible barrier. Difficulty in dealing with a problem is interpreted as a total failure. His cognitive response to a problem or difficulty is likely to be an idea such as "I'm licked," "I'll never be able to do this," or "I'm blocked no matter what I do" . . . In achievement-oriented situations depressed patients are particularly prone to react with a sense of failure [pp. 256–257].

Though the investigations discussed above did not deal with clinical problems per se the pertinence to depression of coping with failure at least is made obvious by Beck. Since internals manifest less perturbance from failure experiences than externals it might be conjectured that an external locus of control and depressiveness are related. A positive correlation has been reported among college students—the more external the student, the more frequently did he report incidents of depression (Abramowitz, 1969). Similarly, Warehime and Woodson (1971) found externals more apt to acknowledge the experiencing of negative effects including depressive feelings than internals, and Goss and Morosko (1970) found a significant correlation between the I–E scale and the depressiveness subscale of the MMPI among three samples of alcoholic outpatients ($r = .24$, n.s. with 62 females; $r = .31$ $p < .01$ with 100 males; and $r = .22, p < .05$ with a second group of 100 males).

The evidence is sketchy. But what evidences do exist indicate that externals are more likely to admit to depressive feelings than are internals. There is something paradoxical about these data however. Why should one feel depressed about that for which he bears little responsibility? Indeed, why should a person experience anxiety if his world is already perceived as unpredictable and uncontrollable?

If anxiety is defined as the state of anticipation of uncertainty, or of the loss of ability to function adaptively, then it would seem more likely to occur among individuals who believe that they exercise some control over their lives. If the world is already an uncertain and uncontrollable place for an external, then one would not imagine him suffering in anticipation of what he already believes to be the case. Despite this appeal to common sense empirical data attest to the opposite. As indicated in Table 2, most investigators report a positive relationship between I–E measures and tests of anxiety—persons holding external control expectancies admit to more experience of anxiety than do those who perceive themselves as internals.

The relationships reported between locus of control and anxiety are rarely very large in magnitude. However, the direction of these relationships is consistent and statistically significant. Only facilitative anxiety is more commonly acknowledged by internals than externals and that measure differs radically from other anxiety scales in that its content focuses upon adaptive responses to stress—the subject is queried as to his responses that should facilitate the likelihood of succeeding at stress-provoking tasks. Obviously this measure is not equivalent to 'other anxiety measures which stress general negative affective responses (Manifest Anxiety Scale) or situation-specific anxiety reactions (Test Anxiety–Death Anxiety).

Investigations in addition to those noted in Table 2 have replicated the finding that external control expectancies are related to admissions of anxiety, depression, and general dysphoria (Burnes, Brown, & Keating, 1971; Goss & Morosko, 1970; Platt & Eisenman, 1968; Powell & Vega, 1972). In short, the evidence counterindicates the seemingly sensible assumption—that negative affect states such as depression and anxiety should be more common among those who are likely to hold themselves responsible for their successes and failures:

TABLE 2

Correlations between Locus of Control and Various Measures of Anxiety

Anxiety Scale	Correlation with I–E	Reference
Alpert–Haber Debilitating Anxiety	$r = .23$ ns, $N = 47$	Butterfield (1964)
Alpert–Haber Facilitating Anxiety	$r = -.68, p < .01, N = 47$	Butterfield (1964)
Alpert–Haber Debilitating Anxiety	$r = .38, p < .05, N = 31$ (M)*	Feather (1967)
Alpert–Haber Debilitating Anxiety	$r = .07$, ns, $N = 53$ (F)**	Feather (1967)
Alpert–Haber Facilitating Anxiety	$r = .07$, ns, $N = 31$ (M)	Feather (1967)
Alpert–Haber Facilitating Anxiety	$r = -.19$, ns, $N = 53$ (F)	Feather (1967)
Test Anxiety	$r = .36, p < .05, N = 46$ (F)	Feather (1967)
Test Anxiety	$r = .13, p < .10, N = 153$ (M)	Feather (1967)
Test Anxiety	$r = .22, p < .01, N = 323$	Ray and Katahn (1968)
Test Anxiety	$r = .21, p < .01, N = 303$	Ray and Katahn (1968)
Manifest Anxiety	$r = .40, p < .01, N = 323$	Ray and Katahn (1968)
Manifest Anxiety	$r = .30, p < .01, N = 303$	Ray and Katahn (1968)
Death Anxiety	$r = .23, p < .05, N = 77$	Tolor and Reznikoff (1967)
Alpert–Haber Debilitating Anxiety	$r = .25, p < .01, N = 142$ (M)	Watson (1967)
Alpert–Haber Debilitating Anxiety	$r = .26, p < .01, N = 506$ (F)	Watson (1967)
Alpert–Haber Facilitating Anxiety	$r = -.15, p < .06, N = 142$ (M)	Watson (1967)
Alpert–Haber Facilitating Anxiety	$r = -.06$, ns, $N = 506$ (F)	Watson (1967)
Manifest Anxiety	$r = .38, p < .01, N = 142$ (M)	Watson (1967)
Manifest Anxiety	$r = .35, p < .01, N = 506$ (F)	Watson (1967)

*(M) = Males.   **(F) = Females.

Phares' hypotheses about this apparent paradox seem eminently reasonable. Internals should have greater difficulty coming to terms with failure if such information was taken to be self-relevant. However, the assumption that internals, like classic hysterics, might become repressive and perceptually avoidant of such information is not convincing. Repression and other such avoidance techniques are more often described as primitive defense mechanisms in contrast to intellectualization, isolation, and sublimation. If locus of control can be said to reflect or be related to social competence or social maturity, it would seem unlikely that an internal locus of control would be associated with a set of less mature or more primitive defense mechanisms.

If we allow ourselves to shift our language style, however, a resolution to the paradoxical findings may be possible. Rather than speaking of defenses, as if everyone must repel or protect themselves against daily travails, it may be more productive to think in terms of coping styles with their more active connotation of struggle and encounter.

In the data derived from correlations between I–E and measures of anxiety, the one reversal, though of a low magnitude, was with facilitative anxiety. If we examine a sample of the items from this scale perhaps some light may be shed upon differences in coping styles that are characteristic of internal and external control-oriented subjects:

> I work most effectively under pressure, as when the task is very important.
> Nervousness while taking a test helps me do better.
>
> When I start a test, nothing is able to distract me [Alpert & Haber, 1960, pp. 213–214].

Each of nine items like the above is either accepted or rejected by the subject and, in at least two studies, internals were more likely to agree with these items than externals. The latter, on the other hand, more often answered the items from the Debilitating Anxiety Scale in the affirmative than did internals. Examples of items from this scale are as follows:

> Nervousness while taking an exam or test hinders me from doing well.
> In a course where I have been doing poorly, my fear of a bad grade cuts down my efficiency.
> During exams or tests, I block on questions to which I know the answers, even though I might remember them as soon as the exam is over.
> I am so tired from worrying about an exam that I found I almost don't care how well I do by the time I start the test [Alpert & Haber, 1960, p. 214].

Agreement with the facilitating anxiety items suggests what we commonly might describe as eagerness or motivation. The student acknowledges the arousing impact of examinations—he becomes nervous and unidirectional. In effect, he appears to be like a man at the starting line before a short dash—keyed up, hyperalert, and, once on the move, totally unconcerned about anything else. The conclusion—whether one wins or loses, becomes secondary to the preparation for and onset of action—the doing that is required in the challenge. In essence, all cognitive—affective activity ceases but for the efforts demanded in the test for the student.

In contrast, debilitating anxiety suggests the reverse. A student who admits to debilitating anxiety is stating that the task, test, or challenge is secondary to his feelings; that is, worry about ends and immediate affective responses are more prominent than the actual process of "doing." Thus, the student becomes "blocked," fearful, and perhaps even disinterested during the very period of testing. In search of an appropriate analogy one is reminded of *Zorba the Greek,* the novel by Nikos Kazantzakis, in which Zorba, the "servant," characteristically attacks challenges with a verve and enthusiasm that bewilders his more ruminative, aristocratic "boss." The latter obsessively worries and fails to accomplish his ambitions whereas Zorba plunges headfirst into one activity after another.

Most pertinent to the immediate discussion is the manner in which Zorba and his boss cope with failure and tragedy. The boss, who is a youthful, philosophically oriented writer, works in fits and starts. Every challenge, whether it be relevant to his work or social life, is engaged in with great hesitancy. In one series of episodes in which a young, attractive widow provides several "come hither" hints, considerable pushing from Zorba is required to encourage his youthful boss to respond. When the young man finally does respond to this and other challenges and eventually becomes witness to tragedy and failure, his response is to become maudlin about the inevitability of man's agony. Zorba, on the other hand, copes buoyantly with failure. He engages in vigorous dancing after suffering catastrophe in his creation of a complex irrigation project. In essence, Zorba teaches his boss that life is full of failure but that one must struggle on, fight, and resurrect oneself in the face of defeat.

To dance a vibrant Greek folk dance after failure is as if to say—"to hell with it—let's get on with living—there are other challenges to be encountered—other windmills with which to tilt." To call such a coping style denial or repressiveness would miss what is perhaps more important—Zorba always remains hopeful about tomorrow. Failure may be incurred in one project, but there are always other tasks, other commitments through which one can rejoin the effortful pursuit of challenge. Today's failures can never be allowed to be overwhelming so as to preclude subsequent efforts. Resurgency, buoyance, and humor would seem to be elements in a coping style which might result in less recall of failures but more preparation and readiness to engage in new challenges.

Though banal by comparison to the literary analogy, the dimensions of facilitative and debilitative anxiety offer a rough approximation to these opposites. In the former there is a sense of excitement in challenge, in the latter, a surrender and a shrinking away from encounter. That these two forms of anxiety are opposites is substantiated by the inverse relationship between them which ranges from $-.37, p < .01$ (Alpert & Haber, 1960) to $-.77, p < .001$ (Butterfield, 1964). In turn, as we indicated previously, internal control expectancies are associated with greater facilitative and less debilitative anxiety, while external control expectancies are associated with the opposite.

In the preceding discussion, the inferred links between locus of control and psychopathology have been based on correlations between self-statements. In essence, we are observing a concurrence of descriptions; those who describe themselves as external in response to the I–E scale also describe themselves as more likely to become depressed or anxious. A second necessary approach is to ascertain whether locus of control is related to pathology as assessed by outside observers.

OBSERVER REPORTS OF PSYCHOPATHOLOGY

Harrow and Ferrante (1969) examined the relationship between locus of control and types of disorders among a sample of upper-middle-class psychiatric patients. Diagnoses were independently established by two experienced clinicians. Patients who were diagnosed as schizophrenic were found to score in a more external direction than all nonschizophrenic patients. This finding replicated an earlier one reported by Cromwell *et al.* (1961).

The patient sample included a group of five patients diagnosed as manic, a type of pathology that is often characterized by the display of grandiosity. Harrow and Ferrante found that the manic patients had a mean score of 4.2 on the I–E scale, which was significantly more internal that the means of all nonmanic patients.

The patients were reassessed on the I–E scale after a six-week period at the clinical facility, with the expectation that there would be some shifts in I–E toward the internal end of the continuum. The first finding of interest was that schizophrenics did not display any increase in internality in contrast to other groups. Depressive patients, on the other hand, did change significantly toward greater internality from the first to the second assessments. Other groups also exhibited shifts toward internality, though to a lesser degree than depressives. The single exception was manic patients who became more external, though the small number of subjects in this diagnostic category precluded statistical significance. In essence, the manics appeared to recover to some degree from an extreme internality (grandiosity) and became more cognizant of their limitations.

While most patient groups decreased in externality and manics increased in externality, only schizophrenics remained relatively the same. The differences between schizophrenics and normals reported by Harrow and Ferrante have been found by other investigators. Cromwell *et al.* (1961) had found schizophrenics to be more external on four measures of the I–E dimension (James–Phares I–E scale, the Bialer–Cromwell scale, and two early prototypes of Rotter's I–E scale).

Other investigators have examined the severity of psychopathology with relevance to the locus of control. Shybut (1968), for instance, found that severely disturbed psychiatric patients who had been rated for severity of psychopathology on the basis of disorders of thought, affect, behavior, and social adjustment were more external than moderately disturbed patients and normal persons. Smith, Pryer, and Distefano (1971) replicated Shybut's findings when severity of emotional impairment was assessed by psychiatric attendants using a behavioral adjustment scale. These investigators found that the severely impaired group was higher in external control expectancies than the mildly impaired group.

Lottman and DeWolfe (1972) have reported that locus of control is related to the process-reactive dimension of schizophrenia. Process schizophrenics exhibit little competence throughout their life span. Usually such patients are disorganized and deviant from earliest recorded observations. Reactive schizophrenics, in contrast, have often been competent until beset by a crisis that leaves them in a state of dismay. Such patients often recover with but little attention from the staff in clinical facilities, a period of respite seeming to suffice in their recovery. Comparing patients on the I–E dimension, process schizophrenics were found to be more external than either reactive schizophrenics or nonschizophrenic control subjects. The latter two groups did not differ from each other.

Palmer (1971) found that not only were psychiatric patients more external on Rotter's I–E scale than nonpsychiatric patients, but that a brief index of competence derived from education, marital status, and occupational level differentiated the most from least external within the psychiatric patient sample. In general, the more internal the patient, the more competent he seemed on the composite index.

Thus far, we may conclude that the admission of anxiety and/or depression, the diagnosis of schizophrenia, and rated severity of personal disorders are related to locus of control. Two studies remain to be discussed, however, in which the relation between pathology and locus of control was in the opposite direction. Goss and Morosko (1970) found that alcoholics scored in a more internal control direction than did the normative samples reported by Rotter (1966). As a secondary point of interest, Goss and Morosko also found within their alcoholic sample positive relationships between locus of control and anxiety, depression, and clinical pathology as assessed by the MMPI; that is, the more external the alcoholic, the more likely he was to respond in a pathological and dysphoric fashion on the MMPI.

Similar results among drug addicts have been reported by Berzins and Ross (1973). They found hospitalized narcotic addicts to be significantly more internal than a rather large sample of university students. Berzins and Ross attempted to explain their findings in terms of experienced self-control via opiates; that is, the addict is said to achieve control over bodily states, anxieties, conflicts, and so on through his use of drugs and, therefore, comes to see himself as more internal perhaps than he had been prior to his addiction. Goss and Morosko contend that alcohol affords the alcoholic a similar means for regulating his feelings, thus resulting in a greater sense of personal control.

Though these paradoxical findings with addicts compel one to search for explanations, the suggestion that momentary impulse control afforded by drugs influences perceived control as assessed by I—E scales seems specious. Alcoholics and drug addicts often are known to deny the fact that they have become dependent upon a drug. It is, consequently, not accidental that an important element in recovery for both alcoholics and drug addicts is the open admission of addiction. Perhaps the more internal responses of these addicts reflect a tendency to deny the very helplessness or slavishness to the drug in question that is so evident to everyone but the addict himself. It may also be this tendency to deny disability that accounts for the lower scores on MMPI indexes of dysphoria among internal addicts, for it seems inconceivable that the addict does not suffer feelings of discontent with his increasingly failure-ridden life.

Rather than allow the impression to remain that the paradoxical findings of Goss and Morosko, and Berzins and Ross with regard to addictions and locus of control are simple and reliable, one should recall that Palmer (1971) found alcoholic patients expressing more external control expectancies than other psychiatric patients. Evidently, the setting, the comparison groups, and possibly the perceived intentions in testing may each contribute to what would seem to be confusing and paradoxical results with addicted subjects.

ADDITIONAL CONTRIBUTIONS

In the studies by Glass and Singer (1972) described in the first chapter, two ego-relevant functions—frustration tolerance, as represented in the decision to persist despite failure, and vigilance, as demanded in proofreading—diminished significantly subsequent to an uncontrollable aversive experience. Likewise, the study of persons who had attempted to commit suicide reported by Melges and Weisz (1971) implicated another ego function, that of time binding, with locus of control. Individuals at the time of their suicide attempt reported a marked loss of awareness of the future as well as a loss of the sense of control.

These ego functions—frustration tolerance, attention, and time binding—are, within psychoanalytic theory, necessary for the maintenance of an integrated, nonpathological personality. Without the efficient and effective operations of the ego a person is more subject to what Freud described as dangers from within or without. As a consequence, he frequently suffers anxiety and erects protective defenses against those anxiety arousing dangers. These defenses, when marked, can so hinder an individual in his goal-directed activity that the resultant helplessness often eventuates in a diagnosis of psychopathology. In fact, Bibring (1953) has contended that the basic mechanism of depression is "the ego's shocking awareness of its helplessness in regard to its aspirations [p. 43]." Bibring views the surcease of activity among depressives as a surrender, not of the goals, but of the pursuit after them, since there is little hope that the pursuit will be effective.

In essence, a weakened ego affords poor prognosis whether through a loss of ability to cope with external challenges or through the *awareness* of one's own weakness and vulnerability. The important point to be made here is that the ego skills noted above, as well as those described in research with information assimilation (DuCette & Wolk, 1973), hypothesis formation (Lefcourt *et al.,* 1973), and information seeking and utility (Davis & Phares, 1967; Phares, 1968) allow us to assert that an internal locus of control and ego strength are likely to coexist within the same persons. In fact, it is conceivable that a perception of internal control may be the phenomenological counterpart to ego strength.

Robert White (1965) had forged a similar link between what he refers to as a "sense of competence" and ego strength in theorizing about the sources of schizophrenia:

> It is important . . . to make allowance for the child's action upon his environment, of the extent to which this action is apt to be successful, and consequently of the confidence he builds up that he can influence his surroundings in desired ways. I use the term sense of competence to describe this, and I think that one's sense of competence is an exceedingly important aspect of self-esteem. We can suppose that a placid child who has always received generous narcissistic supplies, without doing much to earn them, will enjoy an agreeable level of self-esteem, provided he never enters a harsher environment. Self-esteem will be much more substantial, however, in the more active child who feels confident that he can elicit esteem from others by competent performances, by modulated assertiveness, or by personal charm. It is to this second person that we would attribute large ego strength. The first would strike us as too much at the mercy of his surroundings [p. 201].

White subsequently evokes a case for the position that schizophrenic withdrawal derives from a sense of incompetence:

> The schizophrenic has not been able to reach, or at least to maintain, a bearable level of confidence that he can influence others so that they will engage with him on an equal basis, listen to him, understand him, respect him, and give him some of the things he wants from human interaction. Not having the confidence, being constantly forced into submissive, avoidant, and other helpless roles, he has not been able to maintain self-respect, has come to dislike people, and has quite possibly not dared to love anyone [White, 1965, p. 208].

In brief, White contends that a sense of helplessness, as in the present terms, an external control orientation, results in dissatisfaction during social interaction. The person who feels unable to succeed in obtaining desired ends in social interactions will come to a point where those interactions are perceived as aversive. The withdrawal of persons suffering with schizophrenia is then interpreted as avoidance of events that are perceived as offering displeasure.

A second point of some interest discussed by White is his differentiation of "confidence" from self-esteem. Where esteem is defined in terms of the value with which one is held by others or by himself, confidence reflects the expectancy held regarding one's ability to determine his fate through the use of personal skills. This differentiation lends some precision to the locus of control construct. Locus of control and self-esteem are not identical though an internal locus of control should make positive self-esteem a more likely and frequent occurrence.

Another differentiation that may be profitable to consider is that between specific and more generalized expectancies of control. Writers such as Valins and Nisbett (1971) have presented case reports as well as data deriving from elaborate investigation that reveal the salutary effects of shifting patients' causal attributions from internal to external sources. In one case discussed by these authors a man who feared that he was a homosexual was helped to overcome his anxiety and depression by being instructed in how he had misconstrued his own "normal" behaviors. In essence, he was taught that what he had considered abnormalities deriving from personal proclivities were actually natural derivatives from mistaken perceptions of his own body; that is, his difficulties were not "personal" or intrinsic to his character but accidents of perception. This externalization, so to speak, served to lessen the patients' consequent burden of guilt and self-derogation.

In this and other instances discussed by Valins and Nisbett, patients' attributions of causality are the object of therapeutic intervention. Patients are encouraged to shift attributions of cause, sometimes to external, sometimes to internal sources, depending on the circumstances involved in the case at hand. These reattributions are most often of a singular nature—the reconstrual of a drug effect, a symptom, or a reaction. In each case, the misattribution has left the patient believing that he is incapable of dealing with his problems. Therapy, while at the same time encouraging a specific external attribution, can oddly enough serve to reinstate the general sense of being able to act. In social learning terms, shifts in specific expectancies from internal to external can be said to at times encourage the return of confidence or generalized expectancies of internal control. If a person's confidence has faltered, for instance, because he has aspired to goals that are beyond his grasp, then it would seem likely that a lowering of aspirations to a more attainable level may help a person regain faith in his abilities and consequently become more ready to accept other challenges. Thus, specific expectancies or attributions of causality may at times be the obverse of generalized expectancies of control. That discriminations need to be drawn between events that are controllable and those that are not if one is to maintain an internal locus of control appeals to common sense. Its therapeutic value may be appreciated in noting the similarity with the prayer adopted by Alcoholics Anonymous from a sermon by Reinhold Niebuhr: "O God, Give us the serenity to accept what cannot be changed, courage to change what should be changed, and wisdom to distinguish the one from the other."

CONCLUSIONS

While some areas of uncertainty remain, there is good reason to believe, on the basis of the research reviewed, that an external control orientation and abnormal personal functioning are correlated. One ambiguity is the direction of causality— does being ineffective and defensive generate a sense of helplessness; or, the converse, does helplessness generate defensiveness? In all probability the relationship is circular and perpetuated through a vicious circle, though there is little empirical data available to allow for certainty regarding this conjecture.

Second, little information is available as to the coping mechanisms that might allow internals to accept failure with more grace than externals—if forgetting is not the internal's device for specifically coping with failure, then what enables him to persist despite failure? We have guessed at the idea of residual hope among internals: given another opportunity it might be possible to rectify today's errors. The formulation is appealing that internals possess resilience because of a tendency not to take defeats as final. However, there has been little data directly addressed to this issue.

Third, the puzzling results with reported tendencies of internals to forget require empirical and theoretical explanation. We may guess that a busy and involved individual is more likely to forget much that he judges to be trivial each day, and we may assume that it is the internal more than the external who is likely to be the active and engaged person. The empirical data are not available, however, to confirm this hypothesis and thus explain the earlier findings regarding recall differences between internals and externals.

It may be concluded then that there is enough convergence of theoretical and empirical data to support the assumption of correlation between locus of control and psychopathology. What is missing are the factual details that are needed to fill in the gaps related to specific questions of how and why.

8
The Social Antecedents of Locus of Control

INTRODUCTION

In recent years there has been a widening interest in the social origins of locus of control. Though research pertaining to social origins has been on the increase, there is a lack of incisiveness as well as of conclusiveness in the results obtained thus far. Nevertheless, the accumulating evidence from Virginia Crandall's laboratories at the Fel's Research Institute, Mark Stephens' project at Purdue, and from Bonnie Strickland's and Steve Nowicki's concerted efforts at Emory University has revealed some convergences in this area.

In this chapter, research specifically concerned with familial antecedents of locus of control is reviewed, as well as those investigative efforts dealing with a closely related construct, contingency awareness. Much of the research in this area has consisted of correlational findings—comparisons of scores obtained upon measures of locus of control from children, and scores pertaining to child-rearing practices of parents. Despite the apparent consistency of methods, however, there is considerable diversity both in terms of the measurement of locus of control and in terms of how child-rearing methods are inferred. In some cases, children are asked to describe their homes; in others, young adolescents are asked to recall what their homes were like when they had been younger children. In some cases, parents have been observed while they interacted with their children, sometimes in laboratory settings, occasionally in their own homes. In other cases parents have been interviewed about their current parenting methods and attitudes, or in still others have been asked to recall their parenting behaviors which had transpired several years prior to the study in question. Although the variety of approaches is considerable, some relatively consistent findings have been obtained which appeal to common sense. Nevertheless, the reader will quickly grasp the fact that this is still an area ripe for exploration with more questions to be asked than to be answered.

FAMILIAL ORIGINS

Among the earliest studies aimed at an examination of familial determinants of locus of control were those by Chance (1965) and Katkovsky, Crandall, and Good (1967). Chance interviewed the mothers of a select sample of university lab schoolchildren who had completed Crandall's Intellectual Achievement Responsibility Questionnaire (described in Chapter 6). The interview consisted of questions from Schaefer and Bell's (1958) Parent Attitude Research Instrument (PARI—1958) as well as an extended inquiry regarding independence training. The latter inquiry asked the mothers at what ages they expected their children to become capable of performing various acts. Chance found first that the more internal the boys on their total IAR scores, the more likely it was that they had mothers who had expectations for early independence ($r = .47$, $p < .01$). Second, the more educated the mother ($r = .45$, $p < .05$) and the less concern the mother had about controlling her son ($r = .48$, $p < .01$), the more internal her son was likely to score on the IAR questionnaire. As will also be noted in certain other studies these relationships were not obtained with girls. Each of these correlations found with males was statistically insignificant in the female sample.

Katkovsky, Crandall, and Good (1967) conducted their study with the families which were participating in the Fel's Research Institute's Longitudinal Study of Human Development. As in Chance's study, the childrens' locus of control was assessed by the IAR. Given the unique nature of the Fel's Institute, however, it was possible for these investigators to make ratings of parental behavior in the home rather than to rely upon parents' self-reports. Maternal behavior was rated on nine scales of which four proved to be consistently relevant to IAR scores. These were "babying," "general protectiveness," "affectionateness," and "approval instead of criticism."

Babying referred to the extent of parental nurturance, ranging from the imposition of help despite the child's lack of need or desire for it, to a refusal of help when requested. High ratings could be said to reflect "overhelp," low ratings, the withholding of help. Observed maternal babying was found to be strongly related to childrens' IAR scores ($r = .64$, $p < .001$), the more internal children having the more babying mothers. Most particularly, babying was related to I^-, internality for failure ($r = .68$, $p < .001$); the correspondent relationship with I^+, internality for success, was of a lesser though still significant magnitude ($r = .44$, $p < .01$).

General protectiveness was defined as the degree to which parents sheltered their children or exposed them to difficulties, discomforts, and hazards. High ratings indicated sheltering, low ratings, exposure. As with babying, this measure was related to total IAR scores ($r = .64$, $p < .001$) and again, the relationship with I^- ($r = .67$, $p < .001$) was stronger than with I^+ ($r = .49$, $p < .01$).

Affectionateness was defined in terms of warmth and affection versus rejection and hostility. High ratings indicated affection, low, hostility. Total IAR scores correlated significantly with affectionateness ($r = .38$, $p < .05$) though it was largely the I^- half of the IAR which accounted for this relationship. Internality for failure (I^-) was significantly correlated with maternal affectionateness ($r = .46$, $p < .01$), while the relationship with I^+ was insignificant.

The fourth variable, approval versus criticism, pertained to the degree to which parents offered the child praise and approval as opposed to criticism and disapproval. High ratings indicated approval, low ratings, disapproval. Overall IAR scores were related to this dimension ($r = .57$, $p < .001$) indicating that an internal locus of control among children was associated with having parents who were more approving than they were critical. However, unlike the other three ratings of maternal behavior, there were marked sex differences in the obtained relationships. For boys total IAR scores and approval–criticism produced an $r = .63$, $p < .01$ while the equivalent correlation for girls was $.41$, $p < .10$. For boys, approval–criticism was related to I^+ ($r = .45$, $p < .05$) and even more so to I^- ($r = .65$, $p < .001$), whereas for girls the same correlations were $.29$ and $.39$, respectively, both of which are statistically insignificant.

Five other rated behaviors were unrelated to the childrens' IAR scores. These were "restrictiveness of regulations," "severity of punishment," "clarity of policy of regulations and enforcements," "coerciveness of suggestions," and "accelerational attempts."

In addition to the rated home behaviors, the Fel's investigators interviewed and tested both the mothers and fathers of twenty girls and twenty boys. The rather lengthy interviews (2½ hours) were rated for each of four variables:

1. Affection—the amount of overt affection and acceptance that the parents seemed to feel and reported expressing toward their child;

2. Nurturance—the frequency and quality of emotional support and instrumental help given the child by the parent;

3. Dominance—the frequency and intensity of the parents' attempts at influence and control of the child through the establishment and enforcement of rules and regulations;

4. Rejection—the parents' dissatisfaction with the behaviors and personality of their child and the frequency and intensity of their direct criticisms and punishments.

The test completed by the parents was a "Parent Reaction Questionnaire" which assessed the parents' reactions to their child's achievement behaviors in four areas: intellectual, physical skill, mechanical, and artistic achievement.

For the twenty boys there were few if any relationships between their parents' interview ratings and their own IAR scores. Only maternal nurturance even approached being a correlate of IAR scores ($r = .44$, $p < .10$). For girls, however, parental rejection, whether by mothers ($r = -.61$, $p < .01$) or fathers ($r = -.42$, p

< .10), was associated with scores, indicating an external locus of control. Otherwise, the results with the questionnaire were sparse. Paternal negative reactions to childrens' achievement behaviors were inversely related ($r = -.41, p < .01$), while positive reactions tended to be positively associated with total IAR scores ($r = .27, p < .10$). As a point of interest I^+ (internality for success) was more highly related to paternal positive reactions ($r = .35, p < .05$), especially for girls ($r = .59, p < .01$), than the total IAR score.

From home observations, interviews, and questionnaires, then, the results suggest that the "child's beliefs in internal control of reinforcements are related to the degree to which their parents are protective, nurturant, approving, and nonrejecting [Katkovsky, Crandall, & Good, 1967, p. 774]." The maintenance of a supportive, positive relationship between parent and child seems more likely to foster a child's belief in internal control than is a relationship characterized by punishment, rejection, and criticism.

Despite the relative congruence among the findings within this investigation it is noteworthy that the directly observed maternal behavior proved to be more highly related to childrens' IAR scores than were the self-report measures obtained from either parent. This may be due to the self-consciousness aroused among parents when they are questioned about their child-rearing behavior. It is not difficult to imagine the apprehension attendant upon having one's parental adequacy assessed. To be a poor parent can be tantamount to an admission of personal inadequacy, an odious thing for this largely middle-class sample.

Given the limitations of parental self-reports, several investigators who have used such data have corroborated the findings from the more direct observations of parental behavior noted above. Davis and Phares (1969) compared parents' attitudes about child rearing, childrens' reports of parental behavior, and the parents' own locus of control scores with those of their children. The subjects were university age students who were asked to recall their parents' behavior. Rotter's I–E scale was used as the measure of locus of control, and the subjects included in the study were all chosen from the extremes of that dimension. The investigators' guiding hypothesis was that parental restrictiveness and directiveness would inculcate external control expectancies in children whose opportunities to test and experience the consequences of their own behaviors would thereby be curtailed.

Davis and Phares found that extreme internals recalled their parents as having had more positive involvements with them, with less rejection, hostile control, inconsistent discipline, and less withdrawal from them than did extreme externals.

As with the results of the Katkovsky et al. (1967) study, however, independently assessed parental attitudes were found to be largely unrelated to childrens' I–E scores though some interesting interactions with the sex of parent were noted. For instance, fathers of internals were found to have been more indulgent and less protective than had been their spouses, while fathers of externals had been less indulgent and more protective than their wives. Such data suggest that the evaluation of parental interactions and complementarities may be valuable targets for developmental research aimed at elucidating the antecedents of locus of control.

A dissertation by Shore (1967) revealed a pattern of results that was similar to the previously described studies. Shore assessed young children with Bialer's (1961) and Battle and Rotter's (1963) locus of control measures and also obtained the childrens' perceptions of their parents' behavior. As well, parents' self-reported attitudes regarding child rearing were collected. Similar to the previously discussed studies, childrens' locus of control scores were more related to their own descriptions of parental behavior than to their parents' self-reports. Children who perceived their parents as exercising more psychological control, as being less warm and less intrinsically accepting, scored in a more external direction than did children who described their parents in the opposite terms. The magnitude of the relationships between Bialer's locus of control measure and each "perception of parent" variable were as follows: with control ($r = .22$, $p < .01$); with warmth ($r = .43, p < .01$); and with intrinsic acceptance ($r = .46$, $p < .01$). In contrast, only one parental attitude, the father's internality with regard to child rearing per se, was related to Bialer's locus of control scores ($r = .21, p < .01$).

In each of the aforementioned studies, then, parental attitudes regarding child-rearing activities proved to be the least useful information for predicting childrens' locus of control scores. On the other hand, childrens' reports about the warmth and nurturance experienced within their homes was associated with their locus of control scores. Only in the study by Katkovsky et al. (1967) were more direct observations obtained, and these substantiated the reports obtained from the children themselves.

The overall findings, deriving as they do from different age samples, tested with a diversity of locus of control measures and procedures for ascertaining familial experiences, are impressive in their relative consistency. Warmth, supportiveness, and parental encouragement seem to be essential for the development of an internal locus of control. Katkovsky, Crandall, and Good (1967) conjectured that "the security provided by the loving, nonthreatening parent is especially necessary for the child to be able to internalize the responsibility for the negative reinforcements he receives. Conversely, a mother's rejecting, primitive, and dominating behavior encourages her daughter to believe that factors outside her own control are responsible for her rewards in intellectual situations [p. 774]." That parental protectiveness, babying, affection, and approval are related to the development of an internal locus of control indicates that a certain degree of insulation must exist around a child, an insulation from the more aversive experiences, if he is to develop a sense of himself as a causative agent.

It has been suggested by some psychiatric writers (Gardner, 1971) that exposure to traumatic, incapacitating events during early childhood interferes with subsequent reality testing and thus encumbers the learning of basic skills that are components of such important activities as reading. In view of the research pertaining to locus of control, it is possible to add to Gardner's formulation that for a child to develop into a reality-testing adult, one who is aware of his capabilities and limitations, he needs to be reared in a home in which he is relatively sheltered from aversive stimulation that could intimidate him and thus decrease his sense of freedom to explore his milieu. In becoming less exploratory the child would have too constricted a range of experiences from which to discover his particular talents.

THEORETICAL CONSIDERATIONS OF FAMILIAL INFLUENCES

Despite the appeal of the above formulations it would be a marked oversight to ignore a number of questions generated by the data presented thus far. For instance, it may be asked how a "babied" child, one who is overindulged by a hyperresponsive parent, can develop a sense of himself as an active source. It is likely that he receives social reinforcement on a random, if overabundant schedule and therefore may fail to perceive a connection between the occurrence of social reinforcements and his behavior. Alfred Adler described the debilitating effects of pampering and neglect, the former of which corresponds to the common definition of babying. Adler (Ansbacher & Ansbacher, 1956) believed that if children were overindulged or neglected they were equally likely to develop misguided "fictions" about life, and these were frequently in the form of exaggerated feelings of inferiority and helplessness. Pampered children were said to be as thwarted in their development as were neglected children in that they were deprived of the opportunity to learn how to act so as to cause the occurrence of desired outcomes. Both pampered and neglected children, then, through lack of experience with contingent reinforcement, may fail to explore and discover the relationships between acts and outcomes from which beliefs in the order of causal sequences develop. Nonreinforcement for one and indiscriminate reinforcement for the other may result in exaggerated feelings of helplessness, or an extremely external locus of control.

Nurturance and babying, therefore, must be distinguished from the indiscriminate reinforcement implied by the term pampering if the research described in this chapter is to be in harmony with Adler's sensible formulations. Other theorists such as Erich Fromm (1956) offer some help in drawing this distinction. Fromm has discussed "mother love" in a way that clarifies the difference between nurturance that encourages growth and development and the more stultifying aspects of pampering:

> In motherly love ... the relationship between the two persons involved is one of inequality; the child is helpless and dependent on the mother. In order to grow, it must become more and more independent, until he does not need mother any more. Thus the

mother–child relationship is paradoxical and, in a sense tragic. It requires the most intense love on the mother's side, and yet this very love must help the child to grow away from the mother, and to become fully independent. It is easy for any mother to love her child before this process of separation has begun—but it is the task in which most fail, to love the child and at the same time to let it go—and to *want* to let it go [Fromm, 1956, pp. 33–34].

Implied in Fromm's writing pertinent to love is a temporal element that can perhaps define the differences between nurturance, babying, and pampering. Nurturance may be defined as that "intense love" required of parents to enable their children to securely pass through their more vulnerable years. However, this safety and security can become arresting if the parents do not later accept the movement toward independence which safety engenders. A. H. Maslow (1954) has discussed a similar point of view from the perspective of motivation theory. The more expansive motives, curiosity and exploration through the desire for self-actualization, rest upon more basic foundations such as safety and security which are established in the earlier interactions between children and their families. In effect, the warm and protective home that has been found to be associated with the development of an internal locus of control may be described as one where the child is protected in his early years but not squelched; where he is sheltered from the excessive frustrations that can easily occur when a child is young and relatively helpless which, in turn, can engender a more fearful approach to lifes' challenges. That children with an internal locus of control recall their homes as more warm and nurturant settings than do external children may also indicate the acceptance with which their own movement toward independence was greeted. It is a moot point whether persons would recall their homes as warm when their attainment of independence was thwarted by parental overindulgence. The themes of many novels, particularly those of Philip Roth, depict the anguish and anger of children reared in such smothering homes.

RECENT EMPIRICAL ADVANCES

Virginia Crandall (1973) has recently presented a unique set of data concerning familial antecedents of locus of control that are of immediate relevance to the above discussion. Crandall's data are exceptional in that they are comprised of locus of control measures assessed during young adulthood, home observations of maternal behavior during the first ten years, and interviews with subjects during early adolescence.

With this uncommon set of longitudinal data, Crandall expressed considerable surprise to find that "coolness" and "criticality" on the mother's part was often positively associated with internal locus of control scores obtained in young

adulthood. Crandall tempered her response to this reversal of the highly replicated warmth–internality correlations as follows: "I do not mean to imply that
these maternal and familial predictions were composed of cruel rejection, devas-
tating criticism, severe neglect or extreme family pathology, for these are rare in
the Fels sample. I only mean to indicate that from the moderate range of these
behaviors displayed in our sample, those on the "negative" end were associated
with adult internality [Crandall, 1973, p. 11] ."

In addition to this emphasis on the "moderate range of behaviors," Crandall
attributed some of the difference between her results and those of others to the
ages at which locus of control was assessed. As Crandall states:

> It may be that warm, protective, supportive maternal behaviors are necessary for the
> assumption of personal responsibility during childhood, but in the long run, militate
> against internality at maturity. Perhaps internality at later developmental stages is best
> facilitated by some degree of maternal "coolness," criticality, and stress, so that
> offspring were not allowed to rely on overly indulgent affective relationships with their
> mothers, but were forced to learn objective cause–effect contingencies, adjust to them,
> and recognize their own instrumentality in causing those outcomes [Crandall, 1973,
> p. 11].

Among the correlates of locus of control scores independence training proved
to be one of the most reliable. Mothers of internals were more likely to have
"pushed their children toward independence, less often rewarded dependency,
and displayed less intense involvement and contact with them [Crandall, 1973,
p. 12] ." Though the ages at which locus of control and independence training
were most highly related differed between sexes, "nevertheless, the internal
adult has, at some time during childhood, experienced a greater push out of the
nest than the external adult [Crandall, 1973, p. 12] ."

Crandall's conclusions are resonant with the theorizing of Fromm as noted
above. The child is thought to need some degree of security that is afforded by a
nurturant home, but also the freedom to explore his world which is facilitated
by some distance and criticality on the part of his parents:

> In childhood, then, when offspring are most dependent on parental acceptance, it may
> be that the assumption of internality, especially for failure, is expedited if the maternal
> push toward independence is embedded in a warm, supportive maternal–child relation-
> ship. At maturity, however, after the offspring no longer need rely on such maternal
> emotional support, then some previous lack of affectionate behavior and close involve-
> ment in childhood seem to be interpreted as part and parcel of a general maternal assist
> to help them stand on their own feet [Crandall, 1973, pp. 12–13].

When Crandall finally addressed herself to the question of why the maternal
"push from the nest" is so helpful for the development of an internal locus of
control, her response is close to some of the previous discussion in this chapter:
"I would like to suggest that its function is to put the child into more active
intercourse with his physical and social environment so that there is more

opportunity for him to observe the effect of his own behavior, the contingency between his own actions and ensuing events, unmediated by maternal intervention [Crandall, 1973, p. 13] ."

It is evident from the research reported by Crandall and those others we have cited that this area of research, pertaining to the origins of locus of control within the familial setting, is far from complete. Crandall found that by 1973 there had been but 16 studies directed toward an examination of child-rearing practices associated with locus of control, and of the 16 which she included in her bibliography, only four were based on observed parental behavior. Despite these recent efforts it is obvious that this research area is far from being complete with definite and replicated information. However, its importance is not to be underrated. As Mark Stephens (1973) has intimated in his developmental research, it is as plausible to assume that a developing internal locus of control may be a necessary antecedent of the kinds of skill learning that comprise intelligence as it is to assume that intelligence is a determinant of the perception of action–outcome sequences that serve to create an internal locus of control.

CONTINGENCY AWARENESS

An independent construct with some relevance to the literature concerning the origins of locus of control is referred to as "contingency awareness." Although research with contingency awareness is also at a neophyte stage of development, it lends some substance to Stephens' conjecture regarding the circularity between locus of control and intelligence. Though theoretically far removed from the more phenomenological, self-reflecting aspects of locus of control, the molecular perception of links between self-generated activity and resulting environmental changes referred to as contingency awareness seems to be that kind of building block of which locus of control is composed.

John Watson (1966, 1967, 1972; Watson & Ramey, 1972) began his investigations of contingency awareness with direct observations of his own new born child. Given the limited mobility of a two-month-old infant, Watson focused upon eye fixation as an available behavior that was potentially an instrumental act for an infant. Watson extended his arms toward his infant so that his closed fists were about a foot apart and about two feet from the child's eyes. On any one day Watson would open and close one of his hands upon being visually fixated by his infant. The other fist was held immobile regardless of fixation. The occurrence of hand opening, then, was contingent with the infant's looking at the fist that was active on that specific day. Included in this sequence were a day's trials in which the fist that opened did so when the infant fixated on the immobile fist.

What Watson discerned in his 11 days of observation was a growing ease of learning and an accompanying display of positive affect. On the tenth day, he recorded the following observations:

Observation 1: As O [observer] prepared to begin the session S [subject] seemed to show signs of heightened arousal and interest. O therefore decided to begin the game without covering his face. This proved to cause no competing distraction save for an occasional glance now and then (face covering was never again instituted). A game was played to the left and then one to the right. Both of these contingencies were rapidly adjusted to by S.

Notable here was the observation that S no longer seemed to fixate on anything while looking away between fixations on the fist as had been the case in previous games. S now looked away in a continuous arch which returned to the fist.

Observation 2: Given S's apparent mastery of the basic game, O changed the nature of the contingency in an attempt to eliminate a tropistic explanation of S's behavioral adjustments. The new game involved opening and closing the left fist when S fixated on the right fist. At first, S had great difficulty. He distributed his fixations overwhelmingly to the left following each occasional fixation on the right which released the left fist reward. This unadaptive pattern continued until O gave up hope and instituted a "shaping" procedure. The shaping involved having the left fist opening contingent upon fixating to the right of the left fist. The required distance to the right was slowly increased till it coincided with the position of the right fist. The shaping was successful and was completed within 3 to 5 minutes. After a few minutes of stable responding to this "peripheral reward" game, the contingency and reward positions were reversed. The new contingency was learned slowly but required no shaping.

Observation 3: At the beginning of this session, S's apparent delight and "ready attention" was so marked that O noted the feeling of confronting a "sophisticated" subject. Learning was judged to have occurred for both a basic game and a peripheral reward game [Watson, 1966, p. 129].

These observations by Watson merely indicated that a child of but a couple of months of age could learn sequences of events. The infant's eye fixations did not really determine the outcomes though in the child's mind this is not inconceivable. Consequently, Watson would have been hard pressed at this point to conclude that what had been observed indicated at some level that the child had attained an awareness of his causative powers. However, other subsequent observations of his child did lend some credibility to these early speculations:

Observation 4: The subject was observed successfully turning from his stomach to his back on the first occurrence of this act. He was immediately returned to his original prone position. He reestablished a near-successful position and after a bit of a struggle finally rocked past the fulcrum of balance with a swinging movement of his leg. He was again returned to prone position. His speed of turning was faster than previously. Again he was returned to prone position and this time his turning was essentially instantaneous. Again he was replaced and again he turned immediately.

While the sequence of increasing articulation of behavior in Observation No. 3 surely would have done no more than exhilarate O's paternal pride under other circumstances, in the setting of the present case O could not help considering an hypothesis of transfer. If the previous learning games had increased the temporal limits of S's contingency awareness, this could have contributed to the rapid mastery of turning following its initial occurrence.

Observation 5: Following the seemingly complete mastery of a few fixation learning games, S was provided a structured but yet natural contingency situation. A cradle gym was hung in front of S while he was seated in the bouncing chair. The gym was at a distance which would allow S's occasional extended arm movements to contact the gym. Within 20 minutes it appeared that S had clearly established a stable response to striking the gym as its movement ceased. S's eyes were fixed on one of the shiny supporting springs of the gym.

Observation 6: During the preceding six days, fixation games were played and two new variations were introduced. One variant provided a vocal "beep" from O when the designated fist was fixated by S. The other variation provided tactile stimulation to S's legs for correct fist fixation. Learning seemed to be clearly apparent under both of these new modalities of reward.

On this occasion S was seated in his bouncing chair near the table where O and Mrs. O were finishing dinner. Suddenly O noted that S was making gross contortive archings of his body (similar to a wrestler's "bridging" exercise) which when relaxed caused the bouncing chair to react with extensive bouncing. While O worried about the neurological implications of these periodic contortions, Mrs. O proposed that the bouncing effects of this behavior might be rewarding the response. It was true that S had previously "enjoyed" being bounced, but he had never systematically bounced himself. O then proposed that if Mrs. O's hypothesis was correct, S should eventually refine his behavior if the notion of least effort had any merit.

Observation 7: At this time S was systematically bouncing his chair by use of rhythmic arm and leg movements.

Observation 8: Now S was bouncing the chair consistently with use of rhythmic leg movement alone.

These latter observations are presented here simply to illustrate what may be the potential breadth of generalization of a young infant's contingency awareness which has been heightened by experience with artificial contingencies. Obviously the behavior may only be the consequence of a fortunately endowed subject who is enjoying an early advance into Piaget's stage of "secondary circular reactions," which supposedly begins normally around the 4th month of age. Yet given the somewhat unusual history of this particular infant, the proposition of generalization is not untenable [Watson, 1966, pp. 130–131].

Watson's initial observations were quite suggestive if perhaps the overamplifications of a proud parent. However, the rudiments of generalization from the effect of providing early contingent experiences proved compelling enough to have stimulated a series of studies which have added substance to Watson's claim for the importance of contingency awareness. Before describing the research that followed upon Watson's initial observations it is instructive to read his theoretical description of contingency awareness that resulted from his first study:

If an organism is to learn to produce rewards in its natural setting, it needs more than just the behavior which effects the occurrence of rewards. In addition, it must be at least functionally aware of the contingency existing between its behavior and the rewarding stimuli. Without such awareness, the organism would be destined to wait until its stream of behavior emitted the the appropriate response. Moreover, the frequency of emission of that response relative to any other would remain unchanged by either the number of times the reward was elicited or its present desirability. If there exists some awareness that the occurrence of a reward is contingent on the occurrence of a particular available response, the organism is in a position to increase the rate of reward by increasing the rate of the appropriate response. In general then, "contingency awareness" refers to an organisms' functional knowledge that the nature of the stimuli received is sometimes affected by the nature of the behavior the organism is emitting. More specifically, contingency awareness refers to an organism's readiness to react adaptively in a contingency situation and to an organism's sensitivity in the perception of contingency situations when they occur.

In the nicely mechanistic language of computer terminology, one might describe contingency awareness as the organism's tendency to scan or read back its memory records of response output and stimulus input and to select that behavior emitted just prior to the reception of the rewarding stimulus. From this perspective, the following two things might be expected to affect the speed of learning in any contingency situation: (1) the tendency to scan the memory record; and (2) the accuracy of the memory record regarding which behavior preceded the occurrence of the reward. Scanning without memory or memory without scanning would be of no consequence for learning. For the purpose of this discussion, then, both factors will be subsumed in the concept of contingency awareness.

An organism possessing a high degree of contingency awareness is set to learn. It is set to record response-output and stimulus-input data, as well as being set to eventually scan this record whenever a rewarding stimulus is received. In computer terms, the "learning instruction" reads: Find and repeat the response which preceded the reception of the rewarding stimulus [Watson, 1966, pp. 123–124].

In essence, Watson has described in molecular terms the steps in the process of developing a sense of instrumentality. To appropriately repeat a given activity requires an awareness of previous action–outcome sequences; that is, a child must be able to recall his prior actions that caused a given outcome, and to know that those actions were related to those outcomes if he is to repeat the sequence.

To explore these conjectures in a more deliberate and controlled fashion, Watson and his colleagues (Watson & Ramey, 1972) constructed an elaborate device for use in an infant's crib. Electrically controlled mobiles connected to pressure-sensing pillows were constructed so that small head movements upon the pillow activated the mobiles, turning them through a full $90°$ rotation. Infants were enabled by this device to control the rotation of the mobiles which were suspended a foot and a half above their heads. These mobiles were placed in the homes of a number of infants whose mothers monitored event recorders to ascertain the number of times the infant activated the mobile within each daily 10-minute period. Watson and Ramey arranged to have two control groups to compare with the contingency aware sample. In one control group the mobiles rotated but not in response to the infants' head movements, and in the other, the mobile was fixed in position so that in fact it was an "immobile."

Subsequent comparisons of head movements indicated that the two control groups did not differ from one another either at the starting or final sessions. As is evident from Figure 11 there was no marked increase in head movements for either control group. In contrast, the infants for whom the mobile rotated

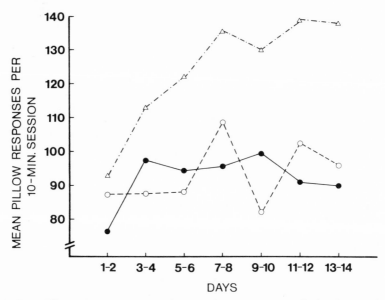

FIGURE 11   Pillow responses per 10-minute session across two weeks for the experimental and two control groups. (△) Contingent, $N = 18$; (●) noncontingent, $N = 11$; (○) stabile, $N = 11$.

contingent upon their own activity increased their head movements from the first to the final trials, as had been anticipated.

These data lend support for Watson's conjectures about the utility of contingency awareness as an explanatory construct. The only difference between the experimental and control groups was in the potential for perceiving contingency between head movements and the activation of the mobile. Therefore it is reasonable to conclude that the infants who produced the increasing number of head movements did so with some level of awareness that these movements were necessary antecedents of the mobile rotation.

Another observation in this study was of some interest. As had Watson's own infant, many of the contingent-aware infants in this larger study exhibited positive affect in response to their causal activity. Although they did not collect such data systematically, Watson and Ramey (1972) noted that "the overall consistency of reports strongly suggests that the experimental infants had a very different 'socio-emotional' reaction to their contingency mobiles than did the controls to their stabiles or non-contingent mobiles ... it appeared that the experimental babies blossomed into smiling and cooing after a few days of exposure to the contingency mobile [p. 222]."

Consequently, Watson and Ramey (1972) concluded that "the infants' reactions to the response-contingent stimulation appear to be of two kinds. The infant reacts intellectually by moving forward to what Piaget . . . describes as the sensory–motor stage of secondary circular reactions. Additionally, he reacts 'socio-emotionally' by displaying vigorous smiling and cooing in the contingency situation [p. 226]."

The provision of contingency experience, then, is regarded as an accelerator of cognitive and affective responses for which Watson has contended (Watson, 1967) the infant is constitutionally prepared. Watson has suggested that the human infant is ordinarily deprived of this acceleration largely because he lacks the behaviors which commonly elicit contingent stimulation from his milieu.

In another study employing the contingent versus noncontingent mobile apparatus, Watson (1972) found replication of the contingency effect and some limits in the infants' ability to discern contingency. If any noncontingent movement occurred in even a small number of trials, then the infants did not perform as predicted. Only with 100% contingent conditions did the increase in movements and affect displays occur as anticipated.

CONCLUSIONS

In John Watson's work with contingency awareness it is possible to see some convergence with locus of control research. Stephens has begun executing a rather ambitious program of research aimed at testing his thesis that an internal locus of control may be an antecedent of much that is referred to as intelligent behavior. Watson too seems prepared to argue that the perception of contingency may be a necessary step in the transition from one developmental stage to another. If these investigators are correct in positing an antecedent role to causal perceptions in generating intelligence reflecting activity, then this area of research focusing upon the origins of locus of control will undoubtedly become a target of increasing activity in the near future.

The research pertaining to familial origins of locus of control seems to indicate that an attentive, responsive, critical, and contingent milieu is a precursor of the development of an internal locus of control. Likewise, as has been described in Chapter 2, the less responsive and less opportune milieu surrounding the poor, the ostracized, and the deprived creates a climate of fatalism and helplessness which is reflected in the scores that individuals obtain on locus of control measures. Lower socioeconomic status, membership in denigrated minority groups, and nonvoluntary quasi-incarceration share commonality in affording minimal contingency between quality of effort and quality of reward, and in generating more external control expectations. The extent of this covariation is evident in recent data reported by Stephens and Delys (1973b), who found not only that preschool children from poor homes were more external than middle-class children, a common enough finding, but within lower-class children, those from homes assessed as above the poverty line were less external than those from homes rated as below the poverty line.

Access to opportunity for contingent responses, then, whether in the home or the larger social milieu, seems to be essential in the development of locus of control. The contingent responsiveness intrinsic to a nurturant but nonsuffocating home and to a responsive and fair milieu seems to be the necessary ingredient for the growth of internal control expectancies that research has uncovered thus far.

9
Changes in the Locus of Control

INTRODUCTION

With the exception of studies such as Glass and Singer's (1972) that concern momentary changes in a person's sense of control, the larger number of investigations discussed thus far contain descriptions of locus of control which make it appear as if it were a stable attribute of persons. Especially in view of the longitudinal data discussed in the preceding chapter the reader may presume that the locus of control construct is viewed as a trait or even as a typology by its users. With subjects commonly classified as internals or externals it could not be argued that this impression is a gross misinterpretation. The difficulty, however, inheres less in the construct itself than in our language which encourages concision on the one hand and an animate subject within each sentence on the other. If group differences are to be described, the more accurate phrasing "those who report (on a given scale) that they perceive events as being largely contingent upon their personal efforts at the present time, as opposed to those who feel more fatalistic about the manner in which outcomes occur," would prove to be too cumbersome to allow for concise expression. Consequently, we are more often confronted with statements suggesting that there are people who *are* internals and others who *are* externals who will, as a result of these *identities,* differ on any number of associated dimensions. It is one purpose of this chapter to attempt to dispel this ready perception of locus of control as a trait, or worse, a typology with all the connotations of intractability and fixedness that those terms imply.

   An individual's locus of control is often inferred from momentary expressions of his sense of causality, which, if solicited at different points of time, may be relatively consistent. However, it must be kept in mind that empirical events such as expressions of causal expectations are but referrants of the locus of control construct and not the construct itself. Locus of control is not a characteristic to be discovered within individuals. It is a construct, a working tool in social learning theory, which allows for an interpretation of remarks made by people in response to questions about causality. The remarks, expressions, and behaviors indicative of beliefs about causality are the events which psychologists observe and test for reliability, and measures such as Rotter's locus of control scale are simply devices created to elicit those expressions of belief. Whether scales such as Rotter's provide adequate opportunity for different persons to give reliable and valid indications of their causal beliefs is a moot point and will be the subject of the following chapter. However, it is to be understood that the responses given to locus of control related questionnaires are not identical to the construct, locus of control, nor perhaps to those phenomenologically real and private thoughts of individuals pertaining to causality. They are but rough approximations of what is believed to be a person's expectancies about control.

Though this brief prolegomenon may strain the reader's patience somewhat it seems to be necessary. For if locus of control were thought to be a trait, consistent and inherent in the person being observed, then evidence regarding change would logically lead one to question the very legitimacy of the locus of control construct. Second, if the construct were to be confused with particular measurement devices, the error variance associated with the imperfections of any psychometric tools would be mistakenly attributed to theoretical limitations of the construct.

However, if the position is adopted that people do not *have* traits such as a locus of control, as if it were a possession, but rather are said to construct interpretations of events, some of which pertain to causality, then it will be less disconcerting to encounter both stability and change in these constructions. People do change their minds or constructions about many things though they as often revert to prior positions, or remain steadfast in their positions despite sometimes overwhelming reasons for changing. It is easier, though, to comprehend both the stability and changes of our constructions if they are regarded as constructions rather than as traits, or other less variable internal attributes.

Likewise, if we remain aware that measurements are but crude approximations of the operations of a construct it will not be as disappointing when changes in these measures or correlations with other theoretically relevant scales do not reach very high magnitudes. No measures will ever account for every last individual quirk possible so that data reflecting upon personality characteristics should not be expected to conform completely to theoretical expectations.

Interest in the processes which can alter one's locus of control is self-evident. In previous chapters, the research discussed revealed that an external locus of control is an impediment to coping with challenges and is associated with negative feelings. The shifting of one's locus of control from an external to a more internal position would seem to be a natural goal for professional psychologists whose aims are often to revive their patients' flagging efforts in pursuit of satisfactions which they have foresaken as hopeless. Erwin Singer (1965) has stated that "... the single proposition which underlies all forms of psychotherapy: the proposition that man is capable of change and capable of bringing this change about himself ... were it not for this inherent optimism, this fundamental confidence in man's ultimate capacity to find his way, psychotherapy as a discipline could not exist, salvation could come about only through divine grace [Singer, 1965, p. 16]."

Granted the professional interest in processes that can alter control expectancies, this chapter will focus upon those events, natural, accidental, and deliberately contrived, that have been explored as potential influences for change of an individual's locus of control.

NATURAL AND ACCIDENTALLY OCCURRING CHANGES

Among the most obvious sources of change in the perception of causality are those associated with age. This author recently had the opportunity of observing this age-related locus of control phenomenon in a discussion with one of his five-year-old daughter's friends. The children had been playing house. Knowing the occasional squabbles that resulted from having to choose between the roles each was to play, I asked my daughter's friend whether she had been "the mommy" or "the baby" on that day. She answered with an immediate and gleeful "the mommy." Asking if she preferred being "the mommy" to "the baby" generated an instantaneous "yes." My quizzical "why" elicited no uncertainty: "Because I get to boss around the baby." "Isn't it as much fun being the baby?" was my naive retort: "No, I don't like to be bossed around."

To be the mommy who directs and determines events in play is obviously preferred by these children, both of whom, incidentally, are the youngest in their respective families. To be young, especially the youngest, means to feel more helpless or external with regard to one's wishes. To acquire a status associated with being older, even in play, then becomes an important source of satisfaction. The easily observed modeling of children after those who are but a few years older than themselves no doubt reflects this valuing of age and the belief that one becomes more capable of obtaining desired satisfactions as one becomes older.

A study by Penk (1969) lends empirical support for this hypothesized relationship between age and locus of control. Penk found chronological age to be positively correlated with internality as assessed by Bialer's locus of control scale ($r = .27$, $p < .01$). This replicated an earlier finding reported by Bialer (1961) in which an $r = .37$, $p < .01$ was obtained between chronological age and locus of control scores. Bialer had also included a measure of mental age derived from the Peabody Picture Vocabulary Test which was found to be related to locus of control scores, $r = .56$, $p < .01$. When Bialer partialled out mental age, the relationship between chronological age and locus of control diminished substantially, leaving a partial $r = .02$. On the other hand, mental age and locus of control remained strongly related $(r = .47, p < .01)$ when chronological age was partialled out.

Penk also included the Peabody test in his battery and likewise found a relationship between mental age and locus of control $(r = .26, p < .01)$. Though Penk did not examine the pattern of his results for effects deriving from partialling procedures, it is reasonable to assume that had he done so he might have obtained results similar to Bialer's. It may therefore be concluded that chronological age per se is not the most salient aspect of maturation with regard to locus of control. Rather, it is the growth of mental age, the extent of vocabulary development, and usage that becomes associated with a sense of being able to determine the shape of one's life.

One other study reveals complementary findings with regard to the passage of time, if not growth, and locus of control. Harvey (1971) found that the longer a person held an administrative position in the upper echelons of government, the more internal he scored on Rotter's locus of control scale. Mean scores on Rotter's scale were 5.4 for administrators holding their statusful positions for 11 years or more and 7.2 for those with but one to ten years experience, the differences being significant $(p < .05)$ even within this generally internal sample.

Though none of these results are "within" or repeated measures as they would have to be to more conclusively test the hypothesis that growth and/or upward mobility in status is associated with shifts in locus of control toward the internal direction, the results do lend some credence to that position.

In addition to these studies which implicate the passage of time and natural changes in status as determinants of change in locus of control, two more serendipitous findings have been reported that reveal the effects of contemporary events upon individuals' perceptions of causality.

Both studies presented locus of control data that were obtained shortly after public events which were pertinent to control expectancies. One report (Gorman, 1968) contained scores upon Rotter's locus of control scale which had been administered on the day following the 1968 Democratic National Convention. Gorman's students were largely supporters of Eugene McCarthy and consequently were highly disappointed and, perhaps, disillusioned with the political process in which they had been so roundly defeated. Though Gorman did not have repeated measures with the locus of control scale, the scores that he obtained on that day were more external $(M = 11.56, t = 5.2, p < .001)$ than the national norms for university students at that time.

More convincing because of the availability of comparison groups was a study by McArthur (1970). Rotter's scale had been administered to a group of Yale undergraduates on the day following a lottery that the United States government conducted to determine draft eligibility for the armed services. Students who were 19 years or older, and therefore affected by the lottery, scored in a more external direction on the locus of control scale than control subjects to whom the scale had been administered prior to the lottery. More important, when those who were favorably affected by the lottery were contrasted with those who were not, the difference in locus of control scores was again significant. Those who were virtually eliminated from draft consideration scored in a more external control direction ($M = 15.5$) than did those who retained the same draft eligibility ($M = 10.22$). Only the favorably affected student differed from the freshman class and control groups while the nonaffected were not different from the comparison samples.

McArthur (1970) described the social context of the lottery so that the meaning of the results became clear:

> Prior to the lottery, most college students probably believed the odds were that they would have to serve in the armed forces after graduation unless they went to jail, left the country, or managed to secure a deferment until they reached the age 26. Hence the lottery clearly brought good luck to individuals ... for this meant that they would probably not have to serve in the army—a real change from their prelottery draft status. On the other hand, the lottery did not as clearly bring bad luck to individuals ... for their draft status remained relatively unchanged in comparison to their prelottery status [p. 317].

Both of these studies convey an important point: locus of control scores shift with relevant environmental events. In these cases the shifts were toward the external whereas the age-related studies indicated change toward the internal end of the continuum. The opposite directions are sensible if one considers the ramifications of the different events. With increases in mental age, chronological age, and length of time in positions facilitating effectiveness, individuals come to perceive themselves as more able to determine the events about them; that is, reinforcements are seen as occurring more in response to personal efforts and/or abilities than they had been previously. Political disappointment, however, especially to those who were so committed as were Eugene McCarthy supporters, creates an impression that the world is unmanageable. Despite political footwork and financial outlays, the Greek tragedy seemed foreordained. In effect, events were beyond control so that participants suffered feelings of impotence that were reflected in more external control scores. Unlike Gorman's study, McArthur's reported upon changes coincident with good fortune. In a sense, those eliminated from the draft could be thought of as enjoying undeserved good fortune. Though opposite in direction the events described in each study were noncontingent upon personal characteristics, and the results were the same. Noncontingent highly valued reinforcements, then, can generate shifts toward a more external locus of control.

Whether these shifts were of any permanence is unanswerable with the data that were presented by these writers. However, it is at least possible to conclude that internal states can be created by environmental occurrences such that individuals will shift in their expressions of the perception of control.

One further study of more clinical relevance illustrates the shifting of locus of control with changing life events. Ronald Smith (1970) examined locus of control scores of a number of clients who appeared at a crisis intervention center in a neuropsychiatric facility. Smith reasoned that since an acute crisis entails feeling temporarily overwhelmed by negative influences, such clients should report that they feel helpless or external. However, as crises become resolved a return to a more internal locus of control should be evident. With a six-week period of treatment focusing upon crisis management Smith found that locus of control scores did decline from the external to the internal significantly. Where crisis clients shifted in locus of control scores from a high at admission ($M =$ 10.08) to a low after six weeks ($M = 7.12$), noncrisis psychiatric admissions remained fairly consistent ($M = 9.63$ and 8.86) for the equivalent test administrations. As one might have expected, then, acute crises are typified by feelings of helplessness at coping with important events. Such helplessness is reflected in locus of control scores which may therefore be taken as a rough barometer of one's sense of being able to cope.

DELIBERATELY CONTRIVED AND BEHAVIORALLY ASSESSED CHANGES

Several investigators, like Smith, have attempted to use locus of control as a criterion or outcome measure for studies pertaining to behavior modification and psychotherapy. In many of these studies locus of control scales have been employed as the primary assessment devices, a procedure that is not without special difficulties. Common sense would suggest that if a person were seeking help he would be ill advised to speak of himself as competent and/or in potential control of his life events. It is more likely that helplessness and a deference toward the helping other would be displayed, since these are more appropriate to help seeking. Likewise, upon discharge from treatment, many clients, either through gratitude to a helpful therapist or through a wish to justify termination of the therapeutic relationship, might wish to express more of an internal orientation.

Fortunately, a few investigators have reported data of a less potentially reactive sort; that is, they have focused upon behavioral changes from which shifts in locus of control have been inferred. Since these are the rarer and more compelling sort of data, these studies will be discussed in somewhat greater detail than those in which scale measures have served as the sole criteria of change.

Research in Educational Settings

Gunars Reimanis (1971) has presented data from three separate investigations in which deliberate attempts were made to alter locus of control among students of different age groups. One study dealt with children from the first and third grades in elementary school. Children for the experiment were selected on the basis of locus of control scores derived from the Battle–Rotter cartoon test, Bialer's scale, and a teacher's rating. Those in the lowest quartile, the most external, were assigned to experimental and control groups.

Reimanis held weekly meetings for three months with the teachers of the experimental group children in which he discussed the means of modifying classroom procedures to encourage children to develop feelings of internal control. These teachers knew the identity of the "experimental children" and as time permitted "gave these children more individual attention with respect to learning about their behavior consequences and consistency in the environment around them [Reimanis, 1971, p. 6]." The later training sessions were devoted to clarifying reinforcement principles by which teachers could identify what was actually reinforcing for each child. Subsequently, they tried to use this informa-tion to point out behavior–effect contingencies to their pupils.

Later, the Battle–Rotter cartoon measure was readministered to the pupils. The experimental group showed a significantly greater change toward internality than did control group children. Furthermore, experimental group children demonstrated more classroom behavior associated with an internal locus of control than did control group children:

> ... the experimental children appeared to know and be interested in what they were doing. They were more a part of their class projects and the teachers could rely on them more. At the beginning of the sessions these children were quite different. Some did not appear to be part of the class. Some showed inappropriate behaviors in the class. Some showed complete lack of interest in classroom activities. And some demanded constant attention from the teachers [Reimanis, 1971, p. 7].

Though these change data were based on small samples, and academic achieve-ment a year later failed to reflect these changes, the initial results proved to be encouraging if uncertain as to cause and permanence.

A second investigation by Reimanis was conducted with groups of community college students. Experimental and control groups were selected from the most external 20% of the student population. One sample received group and another individual counselling sessions aimed at altering the students' locus of control. In the group counselling sessions the counselor confronted students with questions such as "What could you have done about it?" or "What do *you* want to do?" Reimanis also had the counselor attempt to replace external control-reflecting statements ("I am in college because my father wants me to be an optometrist") with internal ones ("I want to learn more about people"). Individual counselling sessions encouraged discussion about vocational and educational goals and raised questions for subjects about the problems they were experiencing with their goal ambitions:

> At each external response the student gave, the counselor interrupted and asked the students to analyze the statement: Why did he say what he said and what could have been done to prevent what happened? The attempt again was to replace an external

response made by the student with an internal one. In addition, the student was encouraged to transfer the internal thoughts to future events. That is, now that he knows what he could have done, what will he do in the future? [Reimanis, 1971, p. 9]."

Experimental group subjects who had received group or individual counselling displayed an increase in internality while no change in locus of control scores were found among control subjects. More importantly, it was noted from the counselling records that most of the experimental group subjects began to model after the counselor's style of questioning and to talk more about their own responsibility for continuing with their education and solving interpersonal problems. In addition, behavioral indications of internality such as taking a new apartment, changing study programs, and seeking out course instructors to ascertain status and needs for improvement were noted among the experimental group subjects.

In one last study Reimanis examined the effects of achievement motivation training courses among college students. In these courses, students participated in game-like situations in which they were able to explore their levels of aspiration and thoughts about achievement. The Rotter scale was administered before, immediately after the training sessions, and again after one to two months had elapsed, and then again seven months later.

Students showed a significant increase in the internal direction following achievement training. Though this increase dissipated somewhat with time, male students retained a significant increase in internality after seven months. However, the equivalent increase that had initially been found among females dissipated in that time. Reimanis conjectured that this may have been due to a lesser concern with academic achievement among females.

Altogether, Reimanis' three studies are suggestive in that they each reveal similar changes in locus of control scores and behaviors relevant to locus of control. However, each of these studies contain weaknesses: small sample sizes, possible experimenter bias effects, and "Hawthorne effects," which prohibit the drawing of clear conclusions. Only in conjunction with other investigations do Reimanis' studies gain credence as demonstrations of changing perceptions of control.

The most ambitious of those studies pertaining to behavioral changes relevant to locus of control is indubitably one by Richard DeCharms whose work focuses upon "personal causation," a similar if somewhat different construct than locus of control. DeCharms (1972) defines personal causation as

... the initiation by an individual of behavior intended to produce a change in his environment. When a person initiates intentional behavior, he experiences himself as having originated the intention and the behavior. He is the locus of causality of the behavior and he is said to be intrinsically motivated. Since he himself is the originator, we refer to the person as an origin.

When something external to the person impels him to behavior, he experiences himself as the instrument of the outside source, and the outside source is the locus of causality.

He is said to be extrinsically motivated. Since the person is impelled from without we refer to him as a pawn. We sometimes talk of people as primarily pawns implying that they more characteristically see themselves as pushed around by outside forces. Conversely, we refer to people as primarily origins implying that they characteristically see themselves as originating their own behavior [DeCharms, 1972, pp. 96–97].

The similarity of DeCharms' origin–pawn dimension with that of internal–external control is evident though there are shades of differences between them. Where locus of control pertains more to the perception of contingencies between actions and outcomes, the origin–pawn dimension focuses more upon the perception of one's self as a subject or object of actions. Nevertheless, the congruities are more salient than are the differences between these constructs, especially with regard to behavioral referents.

In a rather ambitious study, DeCharms (1972) established training programs specifically aimed at encouraging origin behaviors in the schools. His thesis was that to enable a person to behave like an origin the person must be helped "a) to determine realistic goals for himself; b) to know his own strengths and weaknesses; c) to determine concrete action that he can take now that will help him to reach his goals; and d) to consider how he can tell whether he is approaching his goal, that is, whether his action is having the desired effect [DeCharms, 1972, p. 97]."

Black teachers from inner-city schools, with black pupils who were largely from lower-class homes, comprised the samples for this investigation. Experimental group teachers were paid to attend a week-long residential personal causation training course prior to meeting their classes in the fall. Among the course objectives, teachers were instructed in ways to promote origin rather than pawn behavior.

During the year, subsequent to the initial week's training, teachers met regularly with the researchers to design classroom exercises which emphasized self-concepts, achievement motivation, realistic goal setting, and the origin–pawn concept. The latter exercises stressed internal, realistic goal setting, planning, personal responsibility, feelings of personal causation, and self-confidence.

The first result of interest pertained to the way that the children came to see their teachers and the classroom atmosphere. The children were asked a series of questions dealing with the amount of initiative that would be welcome or acceptable within their own classrooms. Children who had trained teachers perceived their classrooms as more encouraging of origin behavior than did children with untrained teachers. These statistically significant differences were obtained from 46 classrooms, 12 children in each class being sampled for their perceptions of classroom atmosphere. With this rather extensive sampling, differences for both the seventh- and eight-grade samples would have to be regarded as indicating considerable success for the training program.

Second, the effects of teachers' training upon student behavior were evaluated in several ways. One set of data derived from content analyses of stories that the children created in each of three grades. The most fascinating aspect of these data is that they were of a longitudinal nature; that is, the same students were assessed during successive grades (fifth through seventh). Childrens' stories were scored for each of six origin—pawn categories: internal goal setting, internal determination of instrumental activity, reality perception, personal responsibility, self-confidence, and internal control, so that scores ranging from 0 to 6 for each of the six stories were obtained each year from the students.

As is evident in Figure 12, origin scores obtained during each grade provided rather striking confirmation of DeCharms' effectiveness in shaping up originlike thought samples. The curve labeled D (bottom right) reflects control group origin scores. Children with untrained teachers did not change throughout the three years in which assessments were made. Curve C (bottom left) reflects origin scores of students who received the benefits of training during the sixth but not during the seventh grade. After training there were decided increases in origin scores which did not decline in the following year though no further training was offered. However, there were also no further increases in origin

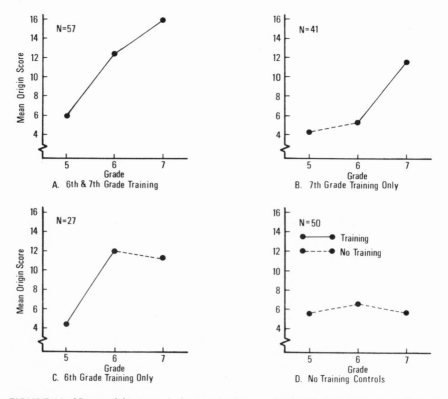

FIGURE 12 Mean origin score before and after motivation training. (From DeCharms, 1972.)

scores. Curve B (top right) illustrates origin scores from a group which received training in the seventh but not during the fifth or sixth grades. Here, it becomes increasingly clear that origin scores change with training. Curve A (top left) shows the cumulative effects of receiving continuous causation training throughout both the sixth and seventh grades. Origin scores increased following each year of training.

These data are fascinating for their detailed, step by step effects. Whenever training was offered origin scores subsequently increased ($p < .001$ in each case). Whenever training was not provided there were no changes in origin scores. Therefore, clear evidence was obtained that personal causation training affected the ways in which children produced imaginative stories. After training, the characters in the stories more often were described as setting their own goals, determining their own instrumental activity, as being more realistic, taking more responsibility for their actions, and as being more self-confident.

In addition to the thought-sample data, DeCharms noted that children in the experimental group became more moderate in their risk-taking behavior during spelling games which entailed choosing more words to spell from lists that were of more appropriate levels of difficulty. Most meaningful for the particular samples under observation, DeCharms found that the usual increasing discrepancy between performances by these inner-city school children and national norms for achievement tests had been arrested by personal causation training; that is, students who underwent the special training offered by teachers who were involved in the personal causation program did not lose ground in achievement scores as they proceeded from the fifth to the seventh grades, whereas control group children exhibited the more typically increasing discrepancy from national norms. Additionally, experimental group children averaged fewer absences than the untrained control group children.

Both DeCharms and Reimanis have shown that behavioral as well as indirect verbal indicators of locus of control can be altered by training programs directed at increasing an individual's sense of personal causation. Each of these investigators conducted his research in educational settings where the concern was largely with achievement-relevant activity, and they both demonstrated the efficacy of educational approaches to changing students' perceptions of causality.

Research in Clinical Settings

Another successful alteration of locus of control was reported in a case study which used completely behavioral criteria. John Masters (1970) discussed a case study in which an "adolescent rebellion" was reduced in severity by therapy which focused upon the reconstruction of causality. The client was a young adolescent who while inebriated had attempted to steal a neighbor's car. That very evening he was apprehended and placed in a detention home where he became agitated and destroyed the furnishings in his cell. From subsequent clinical interviews it was discovered that the youth had been mildly depressed for about a year and that his relationship with his family had become strained to the point where some physical violence occurred. Briefly, the client described himself as "put down" or as a pawn of his parents' wishes. Masters described his therapeutic approach as follows:

> The client's initial perception of the relationship between himself and his parents was one in which the parents were the almighty controllers. He interpreted commands (which may actually have been requests) to mow the lawn or wash the car as infringements on his personal freedom. As these examples were brought out during therapy, it was pointed out that they could be used to his advantage. It was argued that parents learn to reward what they label "good son behavior." This category of behavior includes a large number of distinctly menial chores which by many adolescents are considered degrading. It was then argued that they *are* degrading if one falls into the pattern of performing them blindly or playing the game without being aware of the rules. However, such behaviors could also be performed intentionally and "contingently" as a method for controlling parental behavior.
>
> A series of behaviors were planned in order to demonstrate this hypothesis. On one day it was decided that GB was to mow the lawn without being asked and then to report in detail the effect of his maneuver on his family. The initial report was that his father had difficulty in responding, that he thanked GB, and subsequently he seemed less likely to enter into verbal arguments. It was interpreted to GB that he had become "master" of this game and had turned some tables. Instead of the father's argument, "I pay for the food in this house, you earn your keep by mowing the lawn, etc.," GB had placed his father at a disadvantage by performing some work (equivalent to "I pay for the food . . ."), but had not asked anything in return. However, the father now felt obligated, and GB had very effectively controlled the father's arguing behavior as well by removing a primary inciting stimulus from the father's repetoire. He also had prevented his father from commanding (requesting) him to mow the lawn by his early performance of that task, thus removing an oppressive stimulus from his own environment.
>
> Gradually the above procedure was applied to various good son behaviors such as washing the car or helping out in the family business. The results were consistently positive. Most notable were a reduction in friction between GB and his father and early restoration of the use of the family car which had been denied following the attempted car theft. Tangential to the planning of manipulative good son behaviors were discussions concerning behaviors likely to produce friction. For example, after the use of the car was restored, it was noted that care should be taken in getting home on time, since otherwise the client's father would "automatically" become angry. It was stressed that the father was the foil in this game, since his anger was automatic and unwitting. GB, however, would be the master since he could predict his father's behavior and thus control it by failing to provide the necessary stimuli (coming home late). Getting home on time would also control the father's future behavior concerning late night privileges and future use of the car. The point of the therapeutic argument was always to describe the emission of good son behavior as an effective method for the control of parental behavior [Masters, 1970, p. 215].

Therapy, then, consisted primarily of the verbal reinterpretation of events that otherwise remained the same, which, in turn, caused actual changes in the family environment and in the youth's morale. Feedback some eight months after therapy termination revealed that the client was doing well in school and was in good spirits.

Masters has provided a detailed close-up of youth in crisis, and reports observations that are congruent with the findings of Smith (1970) regarding crisis clients. The youth described by Masters initially reported feeling very much like an object or pawn and, as such, was highly agitated and antagonistic toward his parents, "the controllers." Where Smith noted that crisis resolution was accompanied by a shift toward internality, Masters' purposefully reoriented client exhibited improved morale and coping behavior along with the assumed shift in locus of control which Masters had engineered.

These studies by Reimanis, DeCharms, and Masters are among the few that have not relied entirely upon changes in scale scores to ascertain the effectiveness of therapeutic techniques. The larger number of investigators have used locus of control scales before and after treatments to contrast the shifts in perceptions against control groups that have received either less adequate or no treatments. Though these studies offer some further information regarding methods for changing perceptions of causality, they are subject to more criticisms than the behaviorally focused investigations noted above due to the more likely biasing of responses by their clients.

SCALE-ASSESSED CHANGES

Foulds (1971) found that college students who were engaged in quasi-group therapy which emphasized affective expression, awareness of personal freedom, and responsibility, became more internal on Rotter's locus of control scale. Foulds repeated this experiment with three successive samples with the identical results. A second study by this same investigator and his colleagues (Foulds, Guinan, & Warehime, 1974) reported similar shifts in locus of control scores subsequent to a marathon group session characterized as experiential–gestalt in orientation. Diamond and Shapiro (1973) also found shifts toward internality on Rotter's scale after encounter group experiences in two successive studies, each of which had three experimental and one control group. In each study control groups remained stable in locus of control scores whereas experimental subjects exposed to one or another form of encounter group experience shifted toward internality.

A number of other studies further illuminate the change process. One investigation by Gillis and Jessor (1970) indicated that there were no changes in locus of control scores among hospitalized psychiatric patients who had received brief psychotherapy in comparison to an untreated control group. However, when patients were subsequently classified as to whether or not they had improved in therapy, those adjudged "improved" did demonstrate the predicted shift in their locus of control scores. The simple provision of psychotherapy per se was not found to be an all powerful determinant of change in locus of control scale scores. However, the fact that patients who improved, even with brief psychotherapy, shifted toward internality on locus of control measures encourages the interpretation that such a change in the perception of causality is one element of successful psychotherapy. This conclusion is congruent with Singer's supposition, noted earlier, that the aims of psychotherapy are dependent upon a restoration of self-determination.

Another therapy study (Dua, 1970) compared the effects of a behavioral action-oriented form of therapy with those of a reeducative psychotherapy. The subjects were female university students who came to the university counselling center expressing a concern about their relative inability to relate with others. Subjects were assigned to one of three matched groups, two of which received treatment and one (the control group) in which subjects were placed on a waiting list. One treatment consisted of an action program: "the main emphasis in this method was first to direct the subject to define the interpersonal problem in behavioral terms and then to establish a sequence of specific actions to expand the subject's repetoire of specific behaviors. Tangible actions were suggested by the therapist that would lead the subject to change his actual behavior vis-à-vis the significant other [Dua, 1970, p. 568]." In contrast the reeducative program involved discussions "of the subject's affects and attitudes toward the significant other that might be cultivated or changed and suggestions were made about new concepts and constructs that could be adopted about oneself in relationship to the significant other [Dua, 1970, p. 569]." Control subjects, on the other hand, were placed on a waiting list which delayed treatment for six weeks, the length of time expended in the treatment programs. Rotter's locus of control scale was administered before and after the six-week treatment or waiting periods. Girls who received the action-oriented treatment were found to have increased in internality from their pretreatment locus of control scores; and their posttreatment scale scores were significantly more internal than those of either the reeducative treatment or control subjects. Though girls in the reeducative group shifted somewhat toward internality, they did not differ significantly from the untreated control subjects.

Dua's action-oriented treatment was evidently more effective than his reeducative efforts insofar as changes in the perceptions of control are concerned. This is not too surprising in view of the greater likelihood that cognitive rehearsals and considerations of alternative approaches for interacting with others would have occurred in the action-oriented treatment. Such active exercises bear rough similarity to the kinds of learning tools used in DeCharms's personal causation training.

Nowicki and Barnes (1973) designed a quasi-therapy program in which contingent reinforcement and effectance training were intrinsic elements. The setting was a summer camp provided for deprived inner-city adolescents. The camp was described as being highly structured with emphasis placed upon contingent reinforcement for good or poor performances. Nowicki and Barnes had access to seven successive groups of adolescents ranging in numbers from a low of 27 to a high of 54. In addition, for an eighth week, selected students from the first seven weeks were invited back for another week-long stay at the camp.

While diverse, the camp activities stressed cooperation in the pursuit of group goals, the groups being formed by residence in particular cabins. To accomplish these group achievements a number of conservation projects were undertaken by each group. During these projects campers were socially reinforced for each of their efforts, and at the end of the week actual deeds by individuals were recounted during a public ceremony (Indian council).

All of the campers were administered the Nowicki–Strickland locus of control scale at the beginning and at the end of each week of camp. An overall comparison based on all eight groups was highly significant, indicating that campers expressed some shift toward internality after the week's camping experience. Similar to DeCharms' finding that internality or a sense of being an origin increased cumulatively with additional training toward that end, Nowicki and Barnes found that the youngsters who returned for the extra week's session revealed further shifts toward internality; that is, these children began their extra week with "pretreatment" locus of control scores that were already more internal than any of the "pretreatment" groups; and still they revealed further shifts toward internality. In essence, the more experience with challenge and contingent reinforcement the more internal youngsters were likely to feel.

One other brief set of findings bears discussion here. Most of the therapy studies discussed thus far have been with persons suffering from some deficits for which ameliorists have attempted to provide compensatory training. Writers with interest in the effects of poverty, however, have often spoken of the benefits of becoming a helper as opposed to a "helpee." Frank Riessman (1962), for one, has written of the increase in self-esteem accrued through the helping of others and has demonstrated this phenomenon in therapy programs with certain deprived individuals. As indicated earlier in this chapter evidence has been found that experience in positions that allow effectiveness is associated with greater internality (Harvey, 1971). To help others is, in a very real sense, being effective. Whether in an advisory or counselling capacity, a helper actively seeks to effect another and sometimes can see the explicit results of his efforts. Consequently, it can be hypothesized that the learning of skills which enable a person to become an effective helper should result in a greater sense of internal control.

Two studies bear out this hypothesis. Gottesfeld and Dozier (1966) found that the more experience that "community stimulators" had within their "Massive Economic Neighborhood Development" program in East Harlem, the more internal were their scores on Rotter's locus of control scale. Likewise, Martin and Shepel (1974) found that nurses who requested and subsequently received some training in personal counselling skills exhibited a significant shift toward internality on James' locus of control scale. Their training consisted of three elements: methods for developing helping relationships, for identifying and exploring problems, and for creating plans of action. Role playing was used as an instructional device and the nurses learned how to seek information and to encourage potential help seekers in problem solving. Considerable feedback was available throughout the training procedures so that these nurses were able to assess their own potential effectiveness as counselors.

Given all of this ready, contingent feedback it was no surprise that these nurses exhibited significant shifts toward internality. Accompanying these changes was a correlated increase in scores on a measure designed to assess their likelihood for engaging in facilitative therapeutic interventions. The correlation between posttraining locus of control scores and therapeutic potential ($r = .56, p < .01$) was substantial. As a testament to the effectiveness of that training, this same correlation between locus of control and therapeutic potential before training was negligible and statistically insignificant ($r = .16$).

CONCLUSIONS

Research with both naturally occurring and contrived events has revealed that locus of control scores assessed by scalar and/or behavioral means are susceptible to influence. People change in their customary causal attributions if they encounter experiences that meaningfully alter the contingencies between their acts and perceived outcomes. Though a number of studies have been cited in this chapter, it is obvious that there is considerable room for further information. With the exception of DeCharms' study, for instance, there is little evidence regarding the persistence of observed changes. Though there seems to be considerable consistency in the research on changes in perceptions of control, most of the studies discussed in this chapter could be criticized for potential contaminating effects, whether they be the "Hawthorne effect," experimenter bias, or demand characteristics, each of which can play havoc with designs involving pre- versus posttreatment assessments.

Most critical, perhaps, are the lack of both clear descriptions of the therapeutic approaches and the systematic use of varied criteria in studies that have been conducted thus far. From the research reviewed here it can be concluded that the more action-oriented therapies which stress the learning of and effecting of contingent results seem to be the optimal approaches for changing clients' perceptions of causality. Whether these findings derive from the more clearly explicated task demands or therapist expectancies is not evident at the moment. Though this author is prone to accept the suggestion that behavior modification procedures aimed at increasing contingency awareness are apt to be effective at shifting clients' perceptions of control, there will remain gnawing doubts that can only be assuaged by more extensive and precise research which control for all of those highly contaminating influences that often serve to give us the results that we already anticipate.

10
The Assessment of
Locus of Control

INTRODUCTION

The history of psychology is replete with examples of constructs which have been granted prominence for a brief period of time only to be subsequently discarded and forgotten. However, given the limits of human innovations in describing recurrent phenomena, reinstatements of discarded constructs also occur albeit with differences in language and assumptions. One contributing factor to the common decline of interest in many an adequate construct derives from the mistaken tendency to identify a construct with some singular measurement device: anxiety, a construct that is central to several theoretical positions, becomes defined by Taylor's manifest anxiety scale; repression becomes synonymous with Byrne's repression–sensitization scale; authoritarianism with the California F scale, and so on. In each of these instances a measure became, for many investigators, the definition of the construct, and subjects' scores on those instruments came to be taken as indications of the presence or absence of given traits; that is, individuals scoring high on an anxiety scale are said *to be* highly anxious or *to have* high anxiety.

This tendency to equate constructs with their measurement devices occurs naturally. Typological phrasing is a simple condensation that allows for the communication of considerable information. To say "high anxious subjects will . . . " conveys in highly condensed form the more cumbersome, if accurate, "persons who answered a large proportion of items on a scale that assumedly contains independent items adjudged to be relevant to feelings of apprehension or dread will. . . ." Unfortunately, the casting of scale scores into condensed nominal classifications can have the effect of reifying the relationship between scale scores and constructs which, in turn, can eventually erode interest in the construct when scales lose their utility or fail to afford the precise predictions required for clinical concerns. The awareness that many measurement devices are not meant to be more than crude approximations of an individual's position with respect to a particular construct is easily lost in the pursuit of research employing group statistics with limited criteria.

As most psychology students come to know, there are a host of "error" variables which can contribute to inaccuracy in any measurement device and therefore lessen the utility of that device for measuring actions or cognitions pertinent to a particular construct. Even a ruler, through warpage, may prove to be inaccurate for the precise measurement required in some work; and if such reliable tools as rulers can fail to provide enough precision, then the difficulties inherent in assessing the more metaphorical constructs of personality should be legion. Examiners of personality are faced with variations in habitual expressiveness, cooperativeness, the desire to appear well before others, self-delusions, to say nothing of the problems of task demands, the perceived relationship of test performance to decisions to be made about the person, and so on.

For personality researchers and clinical psychologists these difficulties are of primary importance. Theoretical models and their derivative constructs are invented to describe important personal phenomena that are of focal interest to mental health, psychopathology, and social problems. To test the utility of a given construct crude devices are often created to approximate the relative positions of individuals on that dimension, which is then used to predict behavior on some constructed criterion that bears analogy to a "real-life" relevant phenomenon. If success in predictions abound, the measuring device becomes widely adopted and slowly assumes an identity with the construct that it was devised to assess. However, the construct has meaning within a theoretical network that may have important clinical implications. Given the salience of a particular construct for clinicians, the acceptance of the tool bearing the name of that construct for clinical applications can occur all too easily, and it is in this leap from the demonstration of utility in nomothetic research to clinical purposes that difficulties abound. Clinical prediction requires much data from convergent and divergent sources to raise the probability levels of any prediction. Demonstration research which explores the possible utility of constructs does not require such exactitude. Low magnitude significant results suffice for demonstration purposes but lead to a large number of failures in the exact location of subjects with respect to a given dimension, the very task of the clinician.

Locus of control research, like most research in personality, suffers from these same difficulties and confusions. Rotter's scale, as well as other measures of locus of control, are often taken as if they were equivalents of the construct. While scores on particular measures may be used appropriately for classifying subjects *as if* they were more or less external in orientation, the very process of classification has often led investigators into believing that people *are* internals or externals, or, even worse, "internalizers" or "externalizers"—worse because such terms imply processes that are not immediately relevant to the locus of control construct.

It has been the case in research with many different personality constructs that the failure of assessment devices has caused a decreasing concern with the construct rather than a desire to create better assessment tools. Such misdirected responses to prediction failures may be partially attributed to an only lukewarm interest in the construct since it requires considerable commitment to originate new measures when the old have proven to be fallible. It is therefore noteworthy that there has been considerable effort expended by investigators in the hope of producing newer devices for the assessment of locus of control, attesting to their level of involvement with this construct.

One of the first incentives for creating a new locus of control scale grew from the desire to assess control perceptions among children. Rotter himself, with Battle (Battle & Rotter, 1963), devised one such test that was modeled after Rosenzweig's (1945) Picture Frustration Test. Children were asked to provide answers for characters in a series of cartoons which were later scored for causal attribution. Bialer (1961) was also an early contributor in creating a locus of control scale suitable for children.

It was with the presentation of the Crandalls' Intellectual Achievement Responsibility Questionnaire (Crandall, Katkovsky, & Crandall, 1965), however, that questions regarding assessment aims were first given emphasis. In their paper, Crandall et al. (1965) raised three issues pertinent to their assessment device, one being the question of generalization. While Rotter's scale was originally devised to assess control expectancies in different reinforcement areas (achievement, dominance, affiliation, etc.), factor analysis revealed only one general factor. Consequently, with repeated item analyses the scale was eventually reduced to 23 items that were viewed as being fairly homogeneous. Despite this scalar homogeneity Crandall and his colleagues did not feel that there had been a substantial demonstration that beliefs across reinforcement areas were consistent. In addition, since the Crandalls were engaged in a research program exploring the antecedents of achievement behavior, they were interested in developing a measure that primarily involved achievement reinforcements. With the presentation of their specific reinforcement area locus of control scale, the question of generalizability became salient.

A second problem raised by Crandall et al. (1965) pertains to the sources of external control; that is, in a more general scale such as Rotter's a variety of external coercers are mentioned as potential forces arrayed against self-direction. Luck, fate, and impersonal and personal forces are offered as agents of external control. Crandall et al. (1965) chose to focus upon "significant others," parents, teachers, and peers, as the sources of external control. This restriction upon particular known agents drived from their interests in the natural development of autonomy within the child's world. Most important for the present discussion, however, is the point that Crandall and his co-workers raised, the question as to the particular agents of control, which has become a central question for some investigators.

A third point stressed by Crandall *et al.* (1965) concerns the type of reinforcement—positive versus negative. It was argued by these investigators that attributing responsibility for success can be independent of the way in which individuals interpret their failures. Consequently, the Crandall's locus of control measure, the IAR, was devised to provide separate scores for the response to success and to failure. As have the previous two points, this discrimination in type of outcome has been subsequently explored by other investigators with some success.

In this relatively early paper, then, three issues were raised as being of some importance for investigation with locus of control scales—the generalizability across reinforcement areas, the specification of agents of external control, and the type of reinforcement involved. Each of these aspects has been amplified by certain investigators who have often developed their own devices to correct for perceived shortcomings in other assessment tools.

THE ISSUE OF GENERALIZATION

Generalization across Persons

The first studies directed toward the issue of generalization were less focused upon reinforcement areas than they were with the ways in which minority groups differentiated between themselves and others with regard to locus of control. Gurin, Gurin, Lao, and Beattie (1969) contended that blacks were more likely than whites to draw distinctions in the ways in which outcomes were determined for themselves as opposed to the culture at large. As these authors indicated Rotter's scale contains some items which solicit beliefs about the causes of outcomes for people in general, and some which explicitly refer to the respondent's own life situation. Gurin *et al.* (1969) factor-analyzed Rotter's scale along with several additional items pertaining to personal efficacy and racial ideology. Contrary to Rotter's earlier findings with a primarily white subject sample, Gurin *et al.* (1969) found separate factors for personal as opposed to general causality:

> The items loading on Factors I and II (Control Ideology and Personal Control) are distinguishable in terms of whose success or failure is referred to in the question. The five items with the highest loadings on Factor II are all phrased in the *first person*. The student who consistently chooses the internal alternative on these five items believes that he can control what happens in his own life. He has a strong conviction in his own competence or what we have called a sense of *personal control*. In contrast, only one of the items loading on Factor I explicitly uses the first person. Referring instead to people generally, these items seem to measure the respondent's ideology or general beliefs

about the role of internal and external forces in determining success or failure in the culture at large. Endorsing the internal alternative on these items means rejecting the notion that success follows from luck, the right breaks or knowing the right people, and accepting the traditional Protestant Ethic explanation. Such a person believes that hard work, effort, skill and ability are the important determinants of success in life. We have called this factor a measure of the respondent's *control ideology* [p. 35].

The first factor, "control ideology," includes items 16, 11, 6, 23, 7, 10, 26, 20, and 18 from Rotter's locus of control scale, the order being from the highest to the lower loadings on this factor. The second factor, "personal control," contains items 13, 9, 28, 25, and 15, also in the order of loading magnitude (see Appendix). This isolation of two factors within Rotter's scale was subsequently put to practical test by Gurin *et al.* (1969) and again by Lao (1970). The former investigators found that

> ... students who are strongly internal in the personal sense have higher achievement tests scores, achieve higher grades in college, and perform better on an anagrams task ... in contrast students who are strongly internal in the sense of believing that internal forces are the major determinants of success in the culture at large (ideological) perform *less* well than the more externally oriented students [Gurin *et al.,* 1969, pp. 43–44].

In addition, Gurin *et al.* found that externality on the control ideology factor was associated with readiness to participate in social action and the likelihood of choosing careers that had been atypical for blacks. Lao (1970) found similar patterns of results with personal control predicting academic achievement and control ideology allowing the prediction of social action and civil rights activity.

These findings obtained with black students reveal some limitations in the generalizability of control expectancies. In essence, the black experience provides for some disjunction between the ways in which one perceives one's own and others' experiences; and this is particularly evident in black perspectives upon causality.

Generalization across Reinforcement Areas

A second group of studies pertaining to the generalization issue has focused upon the realm of the reinforcement area. These studies have differentiated between those reinforcements which are directly experienced occurrences in one's private life and those indirect societal events that would more often require a concerted effort by groups. Mirels (1970) was the first of many to report finding two factors in Rotter's scale, one concerning felt mastery over the course of one's life, and the other, the extent to which individuals can exert impact on political institutions. The items from Rotter's scale loading, in order of magnitude on Factor I (felt mastery), were 25, 11, 15, 16, 23, 18, 28, 5, and 10. Factor II, dubbed "system control" by later investigators, contained items 17, 22, 12, and 29 (see Appendix).

The early factor studies did not offer much data of immediate utility. Mirels concluded his own work with a statement calling for studies that might reveal the potential utility of the differentiation suggested by factor analytic results. As will be evident in subsequent discussion, validity research with the different factors has yet to be done. Nevertheless, the work of Gurin *et al.* (1969), Lao (1970), and Mirels (1970) did indicate that there is less than total consistency in the way people perceive events. Locus of control is not to be thought of as a general variable ranging in extremes from total impotence to omnipotence. Rather, the world of events can at least be subdivided for realm of reinforcement on the one hand, and persons for whom attributions are made, on the other.

Subsequent investigators have reported finding similar outcomes from factor-analytic procedures. MacDonald and Tseng (1971), for example, found the same two factors in Rotter's locus of control scale with a male sample as had Mirels. However, MacDonald and Tseng found that while the personal control factor accounted for 15% of the variance obtained with the scale, the social system control factor accounted for a rather small 4 to 5% of the variance. In addition, when these investigators tested a female sample, they found a third factor comprised of items that pertained to social as opposed to achievementlike reinforcements. Items 4, 5, 7, 20, and 26 characterize this separate factor and largely concern the controllability of being liked and/or respected.

Joe and Jahn (1973) have since presented similar data with respect to the personal control and social systems control factors. David Reid and Ed Ware (1973) expanded upon these findings by adding items to the original Rotter scale that should theoretically load upon each of the two factors. These investigators found support for the assumed reliability of these factors. Items pertaining to the control of more distant world affairs loaded on the social system control factor while the more personal items loaded on the personal control factor.

Reid and Ware (1973, 1974) followed up their own investigation with other studies focusing upon multidimensionality within locus of control measures. In one study these investigators found that responses to items pertaining to beliefs about control of impulses, drives, and emotions were independent from either the personal control or social systems control factors of Rotter's locus of control scale. On the other hand, the differences between items of control attributed to self and to others as reported by Gurin *et al.* (1969) were not found. These results, particularly with regard to impulse control, led these researchers to suggest that it was plausible to speak of many independent areas with regard to perceived control.

In support of their contention regarding the utility of a multidimensional approach, Reid and Ware (1974) reported some validity data for the use of separate factors. Personal and Social System Control were used as differential predictors of causal attributions made in response to two situations, one of which lent itself to a personal, and the other to a system control interpretation.

Both Personal and Social System Control were related to causal attributions made about a student who discussed his academic failures on a prepared videotape presentation. Subjects who scored as more internal on either Personal or Social System Control held the student responsible for his failure experience. However, only Social System Control was associated with causal attributions made about a videotaped interview which portrayed a person who had been evicted from his apartment because of a bylaw concerned with the number of occupants allowed in a single residence. At the same time, Reid and Ware found that only social system control was related to political cynicism on the one hand, and political participation on the other. Personal control was not related to the more politically relevant criteria.

Abramowitz (1973) has likewise found that social system control is related to political attitudes whereas personal control is irrelevant to those same political concerns. Given the lack of relationship between personal control and political attitudes Abramowitz concluded, as did Reid and Ware, that considerable predictive power may be lost when the items from both factors are combined without attention being paid to the pertinence of the particular factor to given criteria.

Enough confirmatory evidence has been reported that it is now plausible to speak of a variety of specific areas of control. One investigator has, in fact, isolated four separate factors within Rotter's locus of control scale. Barry Collins (1974; Collins et al., 1973) separated each single forced choice item from Rotter's scale into two items such that the scale became a 46-item Likert scale. In addition to finding that the internal and external items were not highly related to each other, he reported finding four factors after the rotation of his matrix. The factors were labeled "difficulty of world," "unjust world," "predictability–luck," and "political responsiveness." Though Collins acknowledges that each of these factors would be functional equivalents for Rotter's social learning orientation in that each factor would contribute toward or against instrumental activity, Collins asserted that sources of unpredictability in the world are meaningfully distinguishable and relatively uncorrelated.

In response to these data confirming the multidimensional character of locus of control scales, this writer contended at a recent American Psychological Association convention (Lefcourt, 1973) that the discovery of multidimensionality is not surprising, and that the problems attending such discovery are more in the way that subsequent investigators will use diverse factors than as criticisms of the robustness of locus of control as a personality construct.

Locus of control measures may be profitably devised which will differentiate among reinforcement areas or even among persons' perceptions of themselves from others. Investigators such as Oscar Parsons (Parsons & Schneider, 1974) and Walter Mischel (personal communication, 1975) have both found some profit in ascertaining contrasts in the perception of control for one's self and others. The task that would seem to be of the greatest importance now would be in the systematic selection of specific factors to be studied; that is, there would not be much additional benefit from the continuous factor analyzing of given locus of control measures, though recent confirmations of Mirel's work (Viney, 1974) have been of some value. At the present time it would seem most apt that investigators devise specifically aimed locus of control measures for theoretically relevant criteria. Instead of simply rediscovering multidimensionality, it is now appropriate to plan for the assessment of perceived control for specific reinforcement areas.

AGENTS OF EXTERNAL CONTROL

The second point emerging from the original Crandall *et al.* contribution (1965) pertained to the agents of external control. Shortly after the publication of the Crandall article in which the IAR measure was presented, Hersch and Scheibe (1967) published an article concerned with the reliability and validity of Rotter's locus of control measure. In that paper, Hersch and Scheibe reported certain data with adjective check lists which led them to suggest that externality was considerably less homogeneous than was internality. In all, 23 adjectives were checked significantly more often by internals than externals such that internals were found to present a fairly coherent self-portrait. Characteristics assigned to themselves by internals were: "clever, efficient, egotistical, enthusiastic, independent, self-confident, ambitious, assertive, boastful, conceited, conscientious, deliberate, persevering, clear-thinking, dependable, determined, hard headed, industrious, ingenious, insightful, organized, reasonable, and stubborn [Hersch & Scheibe, 1967, p. 612]." In complete contrast, only one adjective was checked more often by externals—"self-pitying."

Hersch and Scheibe (1967) concluded, therefore, that there may be greater diversity in the meaning of externality:

> . . . one may be an external individual because he is in fact physically or intellectually weak in relation to those around him. On the other hand, a person may describe himself as an external because he is in a highly competitive social situation, where the actions of others may have great relevance for the success of his own efforts . . . if a person believes in luck or fate, and if he further believes that these external forces are on his side, he may accurately describe himself as an external. Further, a person may develop feelings of persecution, with or without reason [p. 613].

These differences among external forces bear some similarity to those factors reported by Collins and his colleagues. However, among locus of control investigators, Hannah Levenson has exerted the most effort toward examining the utility of assessing diverse agents of control. In the process she has created a three-factor measure of locus of control which consists of "internality," "control by powerful others," and "control by chance" scales. She has used these scales with some success in predicting specific criteria. For instance, Levenson was able to predict political involvement with her "control by chance" scale while neither "internality" nor "control by powerful others" proved to be relevant to her criterion (Levenson, 1974). Those who believed that events occur in chance fashion were less apt to become politically involved in an antipollution group. As a further point of interest, even among active members of the antipollution group, those who believed in the potency of chance were less knowledgeable about pollution than their less chance-controlled counterparts.

In addition to social action, Levenson has explored the ramifications of her scale measures among institutionalized populations (Levenson, 1973a). Patients diagnosed as psychotic or neurotic were tested at monthly intervals with Levenson's scales. On initial testing, patients scored higher on control by powerful others and chance forces than normal adult samples, though neurotics were much closer to the normal samples than were psychotics. After the second month of hospitalization patients diagnosed as paranoid scored higher on control by powerful others than did all other patients. Finally, with therapy as the intervening occurrence, internality scores shifted toward the internal end of the continuum while control by chance and powerful others remained constant. Levenson (1974) has also found her factors operating differentially among prison inmates. Belief in the control by powerful others was found to be related to the time spent in prison and the frequency of solitary confinement. Inmates who scored high external on the powerful others scale had been disciplined with solitary confinement six times more frequently than inmates who had lower expectations of control by powerful others. Internality and beliefs regarding control by chance were irrelevant to these criteria.

Levenson suggests (1974) that the belief in control by powerful others is similar in nature to the social system control factor of Mirels and others. Her contribution, then, rests in the demonstration of the utility of employing her factored scales as well as in advancing the conception that beliefs in internal control are differentiable from beliefs in the efficacy of chance or powerful others, and that it is possible for all three to coexist independently within individuals.

Support for Levenson's tripartite division of locus of control is evident in a recently reported study by Kleiber, Veldman, and Menaker (1973). These investigators administered each of Rotter's 23 forced choice items as 46 Likert scale items similar to the way that Collins presented the scale to his subjects. Originally paired items were relatively uncorrelated, ranging in magnitude of correlations from .14 to −.47 where high magnitude, negative correlations were expected. Most significantly, a factor analysis of the 46-item scale produced three factors: (1) disbelief in luck or chance; (2) system modifiability; and (3) individual responsibility for failure. Obviously, these three factors bear some similarity to Levenson's chance, powerful others, and internality scales, respectively.

As with specificity of reinforcement, then, there is some evidence in support of the hypothesized specificity of controlling agents. However, it should be evident to the reader that there has not been anything like an exhaustive search for possible external forces that could be evaluated with respect to particular criteria. It would not seem untoward if one were to attempt to delineate a host of possible external forces that might effect outcomes in specific reinforcement areas. An example for such a project is available in Weiner's work pertaining to achievement behavior. As discussed previously, Weiner has described a variety of external and internal forces that can be grouped into categories of stable and unstable sources of control, whether internal or external. It should be possible to do a similar job with reinforcement areas other than achievement, and it may be in such a pursuit that the findings of Levenson and others will reach fruition.

TYPES OF REINFORCEMENT—
POSITIVE VERSUS NEGATIVE

The third area of concern originally discussed in the Crandall paper is that of the type of reinforcement, positive or negative. As was indicated in Chapter 6, research related to achievement activity has revealed that control expectancies relevant to failure may be considered separately from control expectancies for success. Both Crandall et al. (1965) and Mischel et al. (1974) have found that locus of control for failure and for success may be relatively independent from each other and that each may afford the prediction of specified events which the other does not. As noted previously, Mischel, Zeiss, and Zeiss (1974) found that a measure of internality for success experiences was predictive of persistent efforts in activity directed toward the attainment of desired goals, whereas an equivalent measure of internality for failure was better at predicting behavior aimed at avoiding aversive consequences.

A number of other experimenters (Antrobus, 1973; DuCette, Wolk, & Soucar, 1972; Meyer, 1970) have subsequently reported data suggesting that contrasts between the locus of control for success and for failure within the same individuals offer valuable information for behavioral predictions. Most notably, DuCette et al. (1972) reported the data from two studies in which they were able to predict the occurrence of "nonadaptive behavior" from what was referred to as "atypical patterns in locus of control." Lower-class black children in one investigation, and lower IQ children in another study, all of whom were defined as problem children, were found to have rated themselves as more internal for success than they were for failure on Crandall's IAR scale. This was in direct opposition to problem children who were white, had high IQs, and who scored as more internal for failure than for success.

DuCette et al. (1972) concluded that these pattern discrepancies were comprehensible in that problem children were likely to be less sensitive to the environmental feedback that they would probably experience and thus would seem to be less apt to learn from their experiences. As these authors have expressed it,

"In both cases, the problem child by his discrepancy between the assumption of responsibility for positive and negative events has systematically reduced the amount of meaningful feedback he can obtain about himself. The argument being presented . . . is that such reduction in feedback, via these atypical patterns of locus of control, in the long run, might produce maladjusted behavior [DuCette *et al.* 1972, p. 296] ."

While it may be argued that the more normal state of affairs is one in which the locus of control for positive and negative events is symmetrical, the research by DuCette *et al.* (1972) along with that by Crandall *et al.* (1965) and Mischel indicates that much valuable data may emerge from the consideration of separate measures of positive and negative reinforcement control expectancies.

CONCLUSIONS

If one were now to summarize the current status of assessment tools used in the study of locus of control it would be possible to conclude that there is enough evidence to encourage investigators to both continue in their use of existing devices and to develop newer, more criterion-specific measures. If the investigator's purposes are to expand upon the nomological network within which locus of control may operate then devices such as Rotter's scale or Crandall's IAR may suffice despite the failings inherent in each of them. However, if one were seeking to use the construct to make sense of more clinical problems where precision is an issue, then due consideration should be given to the construction of measures that are appropriate to the given problem. Examples of ingenuity in this activity have unfortunately been few to date. One example of such criterion-specific test construction has been reported in a study by Kirscht (1972) who created a perception of control of health scale with which he found some prediction of health-related activity.

If anything is to be derived from the "multidimensionality" research reported thus far, it is that attempts to create such goal-specific devices as Kirscht's deserve encouragement. Demonstrations of multidimensionality need no more repetition. Planful use of specific assessment aims is now in order.

Though other researchers have offered criticism of the locus of control construct and scales referring to ideological biases in the measuring instruments (Fink & Hjelle, 1973; Mirels & Garrett, 1971; Thomas, 1970), and even more critical, to the verbal facility required for comprehending item meanings (Gorsuch, Henighan, & Barnard, 1972), the major criticisms or advice to investigators in this area pertain to the ways in which devices are selected and used. Two recently reported studies have exemplified the wisdom of using the locus of control construct as a moderating variable within a social learning formulation. Wolk and DuCette (1973) found that predicted correlations between achievement motivation and achievement-related behaviors were to be found largely among internal subjects. For instance, the positive correlation expected between a measure of achievement motivation and the preference for intermediate risk was .53 ($p < .05$) among internals in one study and .44 ($p < .01$) in a second study. In contrast, the same correlations among externals were $-.07$ and .10 in each study, respectively. Here, the interaction between value and control expectancy is revealed as a powerful way to enhance the prediction of specified criteria.

A second powerful illustration of the way in which the sophisticated use of Rotter's locus of control measure can lead to good differential predictions is provided in a study recently reported by Murray Naditch (1973). In a paper appropriately entitled "Putting the value back in to expectancy X value theory," Naditch described what he referred to as a means/ends motivational framework. Briefly, Naditch (1973) reasoned that "locus of control should relate to competence only in areas that are important to that person, i.e., areas where competence is related to important goals endorsed by that person [p. 3]." To test this assumption, Naditch obtained self-reported measures of competence in four separate goal areas, assessments of interest in each area, and the locus of control scores from a sample of over 300 public school children. The four competence areas chosen: school achievement, social popularity, sports achievement, and doing things well at home (home achievement) were derived from informal interviews with the school children.

Naditch (1973) found confirmation for the interaction between interest value and locus of control upon reported competence for each of the first three areas among male students:

> Correlations between locus of control and self-reported academic competence, national battery test scores, and grade point average were not statistically significant among male subjects in the bottom two thirds of the distribution of academic interest. For those students in the top third of the interest distribution, i.e., these students for whom academic success represented a highly valued outcome, there were highly significant correlations between locus of control and self-reported academic competence ($r = .378$, $p < .004$), national battery test scores ($r = .398, p < .004$), and grade point average ($r = .400, p < .004$) [Naditch, 1973, p. 11]."

Likewise, with regard to social competence the "correlations between locus of control and self-reported social competence were insignificant for people less interested in social competence, and positively correlated at a significant level for those subjects for whom social competence was more highly valued ($r = .317$, $p < .009$) [Naditch, 1973, p. 11]." With regard to sports competence similar findings were reported. "Correlations between locus of control and sports competence were not significant among people for whom sports competence did not represent a valued outcome, but were significantly correlated among those people high in sports interest ($r = .219, p < .029$) [Naditch, 1973, p. 12]."

Though the findings with female students did not replicate the above, let it suffice to say that at a crude level of assessment, Naditch as well as Wolk and

DuCette have demonstrated the power of using locus of control measures in interaction with assessments of values and interests. In neither case did these authors devise more precise measures of locus of control. It is perhaps in the development of such specific measures, and the use of them within value X expectancy frameworks, that we can anticipate finding improvements in precision and clinical prediction which may save the locus of control construct from the waning of interest that is so common among personality constructs.

11

Current Status
of Theory and Research

OVERVIEW

At this point it would seem appropriate to address ourselves to the question as to what generalizations may be drawn from this literature about the perception of control. It has been a continuous fear among researchers in this area that the dimension of locus of control is simply a euphemism for a "good guys–bad guys" dimension with internality being a substitute for "intelligent, bright, and successful," and externality for "dull, inadequate and failure ridden." In each data realm that we have explored, the findings, in fact, would lead one to believe that the locus of control dimension is an accurate euphemism for such extreme qualities. An intelligent inquiry, however, would include questions as to the reality of beliefs in control, to the problems of grandiosity and delusions of omnipotence, and to the realization that there is much in the way of life experience that must be accepted as inevitable and beyond man's ken. The creation of gods, idols, and other powers attest to man's awareness that he is indeed a limited creature, one that suffers infirmities and threats to his existence that are beyond any reasonable hope of control.

In view of the obvious restraints upon man's self-direction, theorists such as Rotter (1966) have hypothesized that locus of control measures should have a curvilinear relationship with assessments of maladjustment. In other words, individuals who feel themselves to be entirely at the mercy of external circumstances should be no more aberrant in their daily functioning than persons who believe that they are responsible for each and every important event that occurs throughout their lifetime. In the latter case, we approach those pathological processes that would be associated with paranoia, ideas of reference, delusions of grandeur, and so on, while the former would seem to be more relevant to depression, withdrawal, apathy, and retreatism.

Why then, it may be asked, has an internal locus of control most usually been found to be a positive asset and externality a deficit in the research conducted thus far? The answer to this question may be simpler than one might assume. The measurement tools and criterion situations that have been used in most experiments have focused upon events that are largely in the range of controllability for the subjects examined. Achievement reinforcements are often realistically contingent upon the efforts of middle-class North American children. Consequently, internality with regard to achievement-related events is appropriate, and therefore a good prognostic indicator of achievement-facilitating behavior.

No one has yet constructed devices, to this writer's knowledge, which ascertain control beliefs about events that are extremely improbable and commonly believed to be beyond control. Only in the separate factor of "system control," the expression of beliefs about the contingency between individual action and changes in large systems, do we even approach an assessment of less probable event control. With the myth of the democratic New England town meetings still in existence, however, North Americans continue to prize the assumption that individuals can count in remote social events, even if discussions regarding alienation and anomie in large faceless organizations tend to erode that myth. Granted that this myth still persists, if with less general faith in its verisimilitude, system control does not fully qualify as an inappropriate control expectation, especially among privileged middle-class university students. Consequently, an accurate statement of the research to date must take into account a restraining clause regarding the "realism" of control expectancies.

If we define realistic limits and boundaries within which the locus of control variable should have its greatest utility, we may be better able to draw appropriate conclusions from the research literature. To begin with, locus of control is not to be regarded as an omnibus trait similar to "competence" or intelligence which pertains to each and every facet of human endeavor. Rather, it can be more fruitfully defined as a circumscribed self-appraisal pertaining to the degree to which individuals view themselves as having some causal role in determining *specified events.*

Some of the research that we have discussed explored the ramifications of more generalized appraisals or expectancies deriving from cumulative experience with goal striving. Other studies have focused upon more specific self-appraisals relevant to the pursuit of particular goals. Results from both sorts of studies allow us to conclude that the manner in which individuals appraise themselves with regard to causality makes a considerable difference in the ways that many life experiences will be confronted, and we have chosen to describe those ways in terms of "vitality," as will be discussed later in this chapter.

Nevertheless, causal attributions or locus of control represent but one sort of self-appraisal. The concerns, values, or preferences of individuals are of equal importance, as are the expectancies of success or satisfaction. Needless to say, self-appraisals are to be regarded as but one class of information if one hopes to predict behavior with any precision. After all, the opportunities or restrictions present in given situations often obtain greater salience than self-appraisal variables in determining the occurrences of certain behaviors. In short, locus of control is a valuable construct but most especially so when used in concert with other variables of similar relevance to the criteria under investigation.

DEVELOPMENTS IN RESEARCH

Since the writing of this volume was initiated in 1971, there has been an increasing abundance of research relevant to several of the data realms that we reviewed earlier in the book. In this final chapter we shall try to provide some overview with regard to these data realms with at least referencing if not discussion of more recent research as well as of those previously overlooked findings that are germane to these areas.

In addition, we will attempt to delineate some future prospects for research and clinical application that can be derived from the locus of control literature.

Response to Aversive Events

In the introductory chapter research from social, biological, and learning psychology laboratories was used to illustrate a convergent finding—that aversive events vary in their impact depending upon whether the organisms suffering those events can exercise some control over them. Aversive events experienced in a state of helplessness, when individuals do not believe that they can extricate themselves from duress, have more untoward consequences than when those very same events are believed to be controllable.

Further studies by some of the same investigators cited previously have provided support for their earlier conclusions. Glass, Singer, Leonard, Krantz, Cohen, and Cummings (1973), for instance, found that subjective ratings of the painfulness of electric shocks, and the aftereffects of those shocks both diminished when subjects believed that their behavior could effectively reduce shock duration. Glass *et al.* (1973), then, have been able to demonstrate generality of their results across types of aversive stimuli (noise and shock) and types of aftereffect tasks (maze tracing, proofreading, and Stroop test performance). On the other hand, autonomic responses were not affected by the perceived control manipulations.

Other investigators (Bowers, 1968; Houston, 1972) have reported similar findings, that physiological states monitored during experiments in which electric shocks have been administered have not varied with differences in perceived control. At the same time, however, these investigators have found that reported tolerance of shock levels and statements of anxiety shifted in the anticipated directions with controllability. Golin (1974) has likewise reported that uncontrollable shock has more deleterious effects upon complex learning task performances, though this was most particularly so among highly anxious individuals. On the other hand, when controllability of the shocks was emphasized the predicted ameliorative effects upon performance were also found, but largely among the less anxious individuals. Only in two studies thus far, (Geer, Davison, & Gatchel, 1970; Hokanson, DeGood, Forrest, & Brittain, 1971) have physiological responses been reliably altered with the perception of control. In the Hokanson *et al.* (1971) study systolic blood pressure was reduced among those who could instigate rest periods from a shock avoidance task while Geer *et al.* (1970) found lower skin conductance among subjects who believed that they could control shock onset compared to those who did not believe that they could exercise such control.

Further research with "learned helplessness" has also served to substantiate previous results reported by Seligman and his colleagues. Hiroto (1972, 1974) reproduced Seligman's experimental paradigm utilizing inescapable aversive stimuli (loud tones) with human subjects to create "learned helplessness." Hiroto contrasted the subsequent escape—avoidance learning of subjects who had been exposed to pretreatments of either inescapable or controllable aversive stimulation versus those with no pretreatment at all, and in so doing replicated Seligman's learned helplessness phenomenon. Those who received pretreatments of inescapable aversive stimulation exhibited the longest latencies before making avoidance responses in a subsequent aversive task. Hiroto had also obtained locus of control scores (James I—E Scale) from his subjects, and varied the instructions to the criterion task as to the degree of skill required for successful avoidance. In congruence with what one might anticipate if one were extraordinarily optimistic, Hiroto obtained a series of interactions implicating each element of causal perception as an effective determinant of escape and avoidance behavior. Internal subjects were more apt to make avoidance responses and were quicker to learn these responses than were externals regardless of instructions or pretreatments. Likewise, subjects receiving "skill" instructions were superior to those who had been provided "chance" instructions; and externals were more "retarded" with respect to escape latencies than were internals when both had been subjected to the most debilitating inescapable aversive pretreatment.

With the exception of physiological response data, then, most investigators have consistently found substantial support for the hypothesized moderating effect of perceived control upon responses made to aversive stimulation. Occasionally, autonomic response data have been found to buttress these results though with considerably less consistency. A number of investigators interested in the effects of perceived control are presently branching out to consider the responses of persons encountering more "real-life" aversive events such as medical and surgical stress (Cromwell, 1968) and natural disasters such as tornadoes (Sims & Baumann, 1972). However, this work is yet at too exploratory a stage of development to warrant detailed discussion. It is obvious, however, that despite some critical questioning in regard to the generality of statements about perceived control (Averill, 1973; Wortman, 1975) it is fairly safe to conclude that the perception of control has some profound effects upon the manner in which organisms come to grips with adversities.

Response to Social Influences

A number of studies have been published, subsequent to the completion of our fourth chapter, which more or less pertain to the assumed differential responsiveness to social influence of internals and externals. Most relevant to our immediate conclusions are the data reported in a study by Sherman (1973). This investigator conducted an attitude change study contrasting the effects of persuasive messages from without and of self-originated counter-attitudinal essay writing upon subjects grouped according to their scores on Rotter's I–E scale. In accord with our own conclusions regarding self-generated versus externally induced arguments for change, internals were found to have changed more following their writing of counter-attitudinal statements than in response to others' persuasive arguments, while the reverse obtained for the external sample.

In addition to Sherman's study, at least two other groups of investigators have reported data relevant to influence resistance. Snyder and Larson (1972) found that externals were more accepting of extensive personality descriptions derived from minimal test data than were internals. Since this was the very procedure used by Phares, Ritchie, and Davis (1968) which had led those researchers to posit a greater defensiveness regarding failure among internals, the strong possibility exists that the lesser recall by internals of the "personality" interpretations may have derived more from original disbelief than from repressive tendencies. Second, Felton (1971) found that the "Rosenthal effect," the impact of an experimenter's expectancies upon a subject's performance, was maximal when the subject was external and the experimenter internal with regard to locus of control.

Other investigators, while not focusing upon social influence per se, have found, with some reliability, that externals are facilitated and internals distracted and impeded by the presence of social stimuli during various task performances (Baron, Cowan, Ganz, & McDonald, 1974; Baron & Ganz, 1972; Fitz, 1971; Pines, 1974; Pines & Julian, 1972). In an extrapolation from these data, this writer and his colleagues (Lefcourt, Hogg, & Sordoni, 1975), found that internals were more comfortable, as indicated by the frequency of fidgets, when working in isolation than when "feeling observed" by a video camera operating behind a mirror. In direct reverse, externals were found to fidget considerably more often when alone than when observation was made salient.

The overall evidence, then, consistently suggests that externals are more attentive, positively responsive, and facilitated in their task performances by the presence of social cues. Internals, on the other hand, seem to be more resistant to social influences and are, at the least, distracted by social cues as they attempt to cope with various tasks.

Cognitive Activity

Research regarding cognitive activity differences among subjects classified by locus of control scores has not been overly abundant. However, what has been reported tends to substantiate the conclusions discussed in the fifth chapter. Internals have been found to be more verbally fluent than externals (Brecher & Denmark, 1969); internal children have outperformed external children in a recognition of random forms test, particularly when they had to invent their own names or labels for each form. Given supplied labels this difference diminished (Ludwigsen & Rollins, 1971).

Direct support for previous findings relevant to information seeking preparatory to task engagement has been reported by Williams and Stack (1972), and DuCette and Wolk (1973) have found further data which reveal that internals are quicker at extracting cues that will facilitate the making of accurate judgments than are externals, and have found that the former are capable of better recall of performances and are more likely to make use of information for drawing estimates of their subsequent performances than are the latter.

A different sort of data, previously introduced in the fifth chapter, has been reported from this researcher's laboratories that bears upon the cognitive process differences under discussion. In one study (Lefcourt, Sordoni, & Sordoni, 1974), the humor responses noted briefly in an earlier report of that study (Lefcourt, Gronnerud, & McDonald, 1973) were explored in greater detail. Internals were found to have more readily come to terms with the uncertainty inherent in that experimental task, recognizing the discrepancies between the experimenter's stated purposes and actions earlier than externals, and then exhibiting that kind of humor (superiority) that implies a laughing down at one's self for having been deceived. This almost relaxed acceptance of shifting purposes has been replicated in another experiment (Sordoni, 1975) wherein subjects suddenly encountered material indicating that they had been duped—being shown their own photograph within a group of other pictures of "reputed criminals" about whom subjects were registering impressions. Internals again responded in a more relaxed and mirthful manner when confronted with such dissonance than did externals. Perhaps more germane to the cognitive activity realm are the data in another study by Lefcourt, Antrobus, and Hogg (1974). In this study, internals were found to be more able to produce humor, or to be witty, than were externals as they engaged in a series of role-playing interactions.

In each of these humor studies internals were found to more rapidly assimilate the meaning of the ensuing events, to accept those meanings with less rancor than were externals, and to be more able to transform a state of uncertainty to one of humor. Externals, on the other hand, seemed to be less quick to come to terms with their experiences and when grasping the changing demands, exhibited less reactive or productive humor and, at times, appeared dour and irked.

In much of the aforementioned research, then, internals seemed to be more cognitively alert than externals, and more ready to grasp for information that can contribute to the interpretation of and coping with various tasks and situations.

Achievement Behavior

Research that focuses upon the possible relationships between achievement and locus of control has continued unabated though not in a manner that can generate firm conclusions. Some investigators such as Messer (1972) have found support for the simple generality that an internal locus of control (IAR scale) is associated with higher grades and achievement test scores even when IQ and cognitive impulsivity are controlled. More interesting was Messer's finding that boys who assumed responsibility for success (I^+) and girls who assumed responsibility for failure (I^-) were *the* most likely to have obtained higher grades and achievement test scores. This sex-linked difference has been reflected in others' work pertaining to locus of control and achievement. Nowicki (1973), for instance, has found that externality, as assessed by the ANSIE (Nowicki–Strickland I–E scale for adults), was associated with achievement for females ($r = .39$, $p < .05$, $N = 26$) as defined by grade-point average, while internality was related to like achievement for males ($r = -.50$, $p < .02$, $N = 22$). Nowicki subsequently found replication for these data in two other samples of college students with $r = .63$, $p < .01$, $df = 38$ and $.42$, $p < .05$, $df = 26$ for females and $r = -.48$, $p < .01$, $df = 36$ and $-.42$, $p < .05$, $df = 24$ for males.

Consequently, it can be concluded that the verbal expression of perceived causality has different meaning for males and females, at least insofar as scores on current assessment devices are concerned. To further substantiate and add to the confusion regarding sex-related achievement differences, Wolfgang and Potvin (1973) reported that Rotter's I–E scale affords some prediction of classroom participation and academic performance among female elementary school pupils, but not of their male equivalents. However, in this study it was the more *internal* female who had higher average grades and was more likely to be a high classroom participator than was the more external female. In contrast, locus of control was not associated with either achievement criterion for males. These results, while confirming the relevance of sex differences, also reveal the considerable confusion in this area. Different scales used in each study may account for some of the reversals reported. However, the paradoxical findings between sex, locus of control, and achievement behavior seem to be of such perplexity that it will require considerably more theoretical attention toward such issues as socialization of the sexes than it will continued empirical demonstrations to bring some sense of order to this seeming chaos. Nowicki's own discussion of his results begins this very endeavor with discussions of such concepts as the fear of success among females.

More molecular approaches to the area of achievement have revealed some further support for results reported earlier. Gozali, Cleary, Walster, and Gozali (1973), for example, have replicated the findings of Julian and Katz (1968) that pertain to the tendency for internals to require more time before answering difficult than easy problems. Externals, on the other hand, were found to vary less with item difficulty, thus showing lesser adaptability to task demands than internals. These writers suggest, then, that such differences in the appropriateness of responding may be the kind of mediating link that can help to account for relationships between locus of control and achievement.

One other study (Karabenick, 1972) has immediate relevance for this achievement area. Karabenick plotted out curves for the valence of failures and successes against varying probabilities of success among subjects classified according to their locus of control scores. Each subject was asked, in effect, how much satisfaction he attained from success and how much dissatisfaction resulted from failures on given tasks (anagrams and substitution tests) with designated difficulty levels. These data revealed some interesting results with regard to locus of control. When the tasks were construed as being very difficult, internals were found to value success more than did externals, while the reverse was the case for assumedly easy tasks. Similarly, the dissatisfaction resulting from failure was greater for internals than for externals when the task was thought to be easy, while externals seemed to be more discontented with failure experiences than internals when the task was perceived as being difficult. The affective responses of internals, then, were more in accord with what might be expected from realistic, goal-directed, achievement-oriented individuals. On the other hand, the affective responses of externals betray a lesser confidence in their goal pursuits reminiscent of what were referred to as avoidant or irreal responses in the level of aspiration literature. To strongly regret failure at difficult tasks and to be too joyous in response to success at simple tasks suggests an individual with little awareness of or faith in his own potentialities.

Despite the seductive qualities of some of the research pertaining to achievement, the conclusions to be derived from this data realm remain somewhat indefinite. It is quite likely that investigations will continue to be reported at a fast rate in this area. However, until researchers who are concerned with achievement take into account the host of other variables associated with achievement that have been alluded to in the preceding chapters (Naditch, 1973; Weiner, 1972; Wolk & DuCette, 1973), there will not be a sufficient increase in comprehension in this area to justify their continuing efforts. Researchers such as Wolk and DuCette have demonstrated the fact that locus of control can be used as a powerful moderator of achievement value—performance correlates, and Weiner and his associates (1971, 1974) have provided conceptual tools for obtaining more precise attribution statements regarding achievement-relevant experiences. Future investigators would do well to take heed of these writings before plunging into the pursuit of more data that may simply reinforce the sense of confusion that is sometimes almost palpable in the achievement literature.

Psychopathology

Perhaps because the usual assessment tools for measuring locus of control are somewhat inappropriate within clinical settings the number of studies directly pertinent to psychopathology remain limited. However, a few investigators have reported confirmatory data with regard to the assumed relationships between moods and locus of control. Ryckman and Sherman (1973) found a reliable if low-magnitude relationship ($r = -.25, p < .0001, N = 382$) between Rotter's I–E scale and a measure of "feelings of inadequacy," as had Fish and Karabenick (1971) before them; and Gorman (1971) found that external control expectancies were associated with "lower trough mood levels" as ascertained from a month-long diary of moods kept by a sample of junior college students. Another study by Kilpatrick, Dubin, and Marcotte (1974) also made use of self-reported moods elicited from students who were engaged in each of four years of medical school. In this study, medical students characterized as internals exhibited less mood disturbances on the POMS (Profile of Mood States—McNair, Lorr, & Droppleman, 1971) than their more external counterparts. Internals rated themselves as "less tense, anxious, depressed, hostile, fatigued, and confused" than did externals in each of the four classes (freshman—senior). One interaction was of some interest. Though students generally indicated that the second and third years of medical school were the most stressful and unsettling, internals and externals differed dramatically with regard to their ratings of "vigor—activity" in those middle years. Where no differences with regard to vigor were obtained in the freshman and senior year groups, externals reported feeling markedly less vigor than did internals in the second- and third-year groups. Internals remained buoyant in those more stressful years in contrast to externals who described a marked decline in this instrumentality facilitating mood.

In effect, each of the above studies confirms the general finding that an external locus of control is associated with a predominance of negative affective experiences as was dramatically demonstrated in the Melges and Weisz (1971) study concerned with suicide. Kilpatrick et al. (1974) add to this literature in their demonstration that internals maintain their mood of "vigorousness, exuberance, and high energy" in the face of what must be continuously uncertain and arousing circumstances. Partially in response to these data, this author has also investigated the mood—locus of control association with the POMS measure and has replicated the findings of Kilpatrick et al. (1974) in part with undergraduate students. Both tension and depression proved to be more common self-attributes of externals than they were of internals, whereas vigor was more characteristic of the latter than of the former. If tension and depression can be construed as debilitating and vigor as a mood-facilitating instrumental activity, then such results with mood measures indicate that internals are less likely to succumb to demanding circumstances and to remain active in the confrontation with challenges. It is interesting to note, in this regard, that investigators who have focused their efforts toward making depression comprehensible (Ferster, 1973; Lewinsohn, 1972) have often spoken of the decline of instrumental activity or the breakdown of behavior—outcome connections as a major source of depressive episodes.

One last study that requires inclusion in this updating and summary is that by Miller and Seligman (1973). Though these investigators found no data of significance with locus of control measures, they did find that depressed individuals, defined by scores on Beck's Depression Inventory (Beck, 1967), failed to respond to their reinforcement experiences in "skill tasks" as sources of information for deriving estimates of their future chances of success. Consequently, Miller and Seligman concluded that one important aspect of depression is the perception of outcomes as being noncontingent or independent of actions. In this study, the criterion of noncontingency was predicted by scores on a measure of depression. This result complements the findings noted previously, that an external locus of control, the perception of noncontingency between acts and outcomes, predicts to expressions of depression and lack of vigor.

As this review makes evident, locus of control has been most commonly investigated with respect to depressive kinds of disorders. As yet other forms of behavioral pathology have not received the same degree of attention vis-à-vis the locus of control construct. It is not accidental, however, that depression should be the disorder most commonly associated with fatalism or expectancies of external control since the very terms despair and hopelessness are near cognates of fatalism. In the research literature substantiation has thus been found for a common sense linkage that the wisdom inherent in our language would suggest.

ANTECEDENTS, CHANGE, AND ASSESSMENT

The remaining three chapters, because of their greater recency, require less updating than do the prior chapters. Little data have been reported that would alter conclusions regarding the effectiveness of a responsive but nonsuffocating home atmosphere for engendering internal control expectancies in children. As yet, the recently reported longitudinal "discoveries" by Crandall, and Watson's contingency awareness research have barely been assimilated by researchers concerned with the sources of control expectancies.

One study has been reported by Levenson (1973b) which does represent an advance in this area. Levenson explored the "perceived parental antecedents," the retrospective accounts of young adults with each of her control-related measures (internal control, powerful others, and chance scales). Briefly, she found that for males internality was associated with perceived maternal instrumentality whereas for females internality was negatively related to maternal protectiveness. Control by powerful others was associated with reports of parents using more punishing and controlling types of behaviors; and control by chance was related to a perception of parents as having unpredictable standards.

The content of Levenson's findings, while sensible and interesting, are of less immediate importance than is the demonstration of a pursuit of information about specific facets of causal perceptions. With more differentiated causal concepts as criteria, research involving social and parental antecedents may produce higher magnitude relationships than have previous investigations which have focused upon the more generalized locus of control measures.

In research pertaining to changes in the perceptions of control there have also been a few recent advances which make use of Weiner's attributional refinements. Dweck and Reppucci (1973) initially found that Seligman's "learned helplessness phenomenon" could be engendered among children when, with a particular experimenter, they suffered continuous unavoidable failures. A number of these children subsequently failed to complete problems administered by the "failure experimenter" when those problems were, in fact, solvable. These failures occurred despite the fact that the children had solved almost identical problems previously administered by the "success experimenter." In addition, these investigators found that the greatest decrements in performance following failure were evident among children who were assessed as externals on the IAR questionnaire, or among those who, when they did attribute the cause of their outcomes to internal characteristics, did so to ability (stable–internal) rather than to expenditure of effort (variable–internal). Children who persisted despite failures were more apt to attribute outcomes to effort than they were to the stable characteristic, ability.

Given the positive value of persistence in the face of failures for achievement behavior, Dweck (1975) subsequently experimented with a process of causal reattribution in an attempt to produce that very persistence which had been associated with the tendency to attribute cause for outcomes to effort.

Dweck obtained a small sample of children who were adjudged by school personnel to be helpless, in the sense of being passive before challenges which they were capable of achieving. These children were found to differ considerably from "nonhelpless" children both with regard to IAR questionnaire scores and measures of failure avoidance, thus validating the judgments offered by the teachers and guidance counselors. Dweck then contrasted two training procedures that were designed to alter the failure-avoidance behavior expected of helpless children. One procedure was designed to teach the children to assume responsibility for their failures, but with attributions made to effort rather than ability. The other procedure, on the other hand, provided an "enriched diet" of success experiences. When Dweck interpolated failure experiences among the long sequences of successes enjoyed by all of the children she found that the children who experienced the success diet deteriorated rapidly in their performances. However, those who had received reattribution training maintained or even improved in their performance following failures and were found to continue verbalizing the fact that insufficient effort rather than inability was responsible for failures.

Dweck's investigation is compelling for several reasons. First of all, her subjects were real failure-avoidant children whose behavior coincided with that of children who had experienced experimentally induced learned helplessness in the first study. Perceiving that external attributions were associated with a ready capitulation to failure, and that internal stable attributions also serve to perpetuate avoidant behavior, Dweck chose to train her "helpless children" to perceive internal variable characteristics as being responsible for their failures. While this approach drew appropriately upon the results of her first study, it also capitalized upon the element of hope, or a belief in the possibility of change that is implied in variable internal attributions. Failure due to inability is more apt to be thought of as insurmountable whereas ascribed lack of effort invites challenge for change. Dweck's success in generating persistence among helpless children attests to the wisdom of her choice. Within this study, then, there was a sophisticated selection of the criterion variable (persistence despite failure) and a wise use of a more refined subcategory of perceived causality. Hopefully, more therapy-relevant research will make use of similar refinements.

With regard to assessment (Chapter 10) there is less to add though it should be noted that Julian Rotter (1975) has written an article which presents a viewpoint rather similar to that expressed in our tenth chapter. One omission from our previous review pertains to the relationship of locus of control scores with social desirability. While it has been noted that the need for approval bares little relationship with locus of control measures, several investigators (Cone, 1971; Hjelle, 1971; Joe, 1972) have found significant relationships between Rotter's I–E scale and Edward's Social Desirability Scale (SD). Among clinic outpatients, prisoners, alcoholic inpatients, and career planning youths, SD and I–E measures were significantly related to one another with rs varying between $-.29$ to $-.70$ (Cone, 1971). Though the magnitude of the relationships vary considerably between settings and samples, the overall results indicate that there is a tendency for people to evaluate internal attributions in a more positive light than they do external attributions, a fact that should be known by researchers who plan to use the more reactive measures of locus of control such as Rotter's I–E scale.

CONCLUSIONS AND PROSPECTS

If we summarize the bulk of the findings discussed in this book, within the limits acknowledged above, it would be fair to conclude that internal control expectancies about personally important events, that are to some reasonable degree controllable, will be related to signs of vitality—affective and cognitive activity which indicates an active grappling with those self-defined important events. Where fatalism or external control beliefs are associated with apathy and withdrawal, the holding of internal control expectancies presages a connection between an individual's desires and his subsequent actions. As such, locus of control can be viewed as a mediator of involved commitment in life pursuits. If one feels helpless to effect important events, then resignation or at least benign indifference should become evident with fewer signs of concern, involvement, and vitality.

The best example with which this writer is familiar that can convey the essence of our concluding generality linking perceived causation and vitality is contained in a section of James Fenimore Cooper's classic novel, *The Deerslayer*. Cooper (1910) described a scene in which Deerslayer is bound tightly to a tree and surrounded by an execution squad of his enemies, the Huron Indians. The warriors of this tribe cut at Deerslayer's flesh repeatedly with varying accuracy and ferocity so that his torso becomes gashed and oozing with blood. Despite their intimidating tortures, Deerslayer reveals progressively *less* rather than *more* fear that the warriors had anticipated. His stoical countenance proves to be a disappointment if not a threat to the assumed fearsomeness of his captors. Consequently, his assailants pauses in their execution ceremony and begin discussing the situation:

> I see how it is," he said. "We have been like the palefaces when they fasten their doors at night, out of fear of the red-man. They use so many bars, that the fire comes and burns them before they can get out. We have bound the Deerslayer too tight; the thongs keep his limbs from shaking, and his eyes from shutting. Loosen him; let us see what his own body is really made of.
>
> The proposal of the chief found instant favor; and several hands were immediately at work cutting and tearing the ropes of bark from the body of our hero. In half a minute Deerslayer stood as free from bonds as when, an hour before, he had commenced his flight on the side of the mountain. Some little time was necessary that he should recover the use of his limbs, the circulation of the blood having been checked by the tightness of the ligatures; and this was accorded to him. It is seldom men think of death in the pride of their health and strength. So it was with Deerslayer. Having been helplessly bound, and, as he had every reason to suppose, so lately on the very verge of the other world, to find himself so unexpectedly liberated, in possession of his strength, and with a full command of limb, acted on him like a sudden restoration to life, reanimating hopes that he had once absolutely abandoned. From that instant all his plans changed. In this he simply obeyed a law of nature; for while we have wished to represent our hero as being resigned to his fate, it has been far from our intention to represent him as anxious to die. From the instant that his buoyancy of feeling revived, his thoughts were keenly bent on the various projects that presented themselves as modes of evading the designs of his enemies; and he again became the quick-witted, ingenious, and determined woodsman, alive to all his own powers and resources [Cooper, 1910, pp. 349–350].

In Cooper's discerning commentary we can see the details of our concluding generalization. When one believes that hope is possible, that there is opportunity to act in one's own behest, then he becomes more "determined" and "alive to all his own powers and resources," or in a word, vital.

Until this point we have been largely in the process of summarizing and looking backward at the sequence of research discoveries that have led us to recognize the value of the locus of control construct. We have concluded that the perception of control is a useful construct which allows us to comprehend a diversity of human characteristics that we have subsumed with the term vitality. Now our attention may be drawn to future prospects.

Throughout this book, and particularly in the chapter pertaining to assessment, cautionary notes were sounded regarding the ways in which the locus of control construct could be abused and misunderstood. This writer has occasionally been queried as to the advisability of using one or another measure of locus of control by investigators wishing to include the variable in their studies. When he has subsequently inquired as to the particular questions to be asked, or criteria to be assessed, as a determinant of the appropriateness of choosing a device, the response has occasionally been one of puzzlement rather than gratitude. Such encounters have often left this writer pondering about the more general problems of psychology, a field where fads and fashions and the demands for quick and easy results abound. What is often lacking in proposals and in many completed investigations with reference to locus of control are detailed plans or at least a clear, well-articulated rationale for the inclusion of this variable. More unfortunate is the occasional lack of even common sense in the use of locus of control measures in some studies.

As commonsensical guidelines the following suggestions are offered in the hope that the perception of the control variable will continue to be used fruitfully without the discouragement that may result from its misuse.

First, the locus of control construct per se should not be expected to account for a lion's share of the variance in most situations. The perception of control is but a single expectancy construct. Other interacting variables of equal importance, if not more, include the value of the reinforcements in question, as well as the expectancy that one will obtain that desired reinforcement, whether by one's own or external forces.

Second, people are not totally internals or externals. The terms are used as expressive shortcuts and are not meant to imply that perception of control is a trait or a typology. The perception of control is a process, the exercise of an expectancy regarding causation; and the terms internal and external control depict an individual's more common tendencies to expect events to be contingent or noncontingent upon their actions.

Third, if one wishes to use the perception of control as a powerful predictor, then it will most always be profitable to design one's own assessment devices for the criterion of interest. This is similar to stating that people are not so much to be characterized as internals and externals as they may be said to hold internal and external control expectancies about different aspects of their lives. If one were concerned about a particular aspect, such as the ability to maintain close intimate relationships, then the perception of control of love and affection responses would be more salient than would control expectancies pertaining to achievement. To summarize this point, one would always do well to add an "of what" after the phrase perception of control.

Fourth, we must always bear in mind that there are many confounding elements in the term control. It has been contended by some that the construct was originally misnamed, that control was never the central issue, but rather that contingency was at the core of the construct. The term control connotes successful manipulation. The construct, locus of control, on the other hand, focuses upon the perceived contingency of events, whether they be positive or negative outcomes. In addition to the confounds associated with the term control, other writers such as Ivan Steiner (1970) have drawn distinctions between "outcome" and "decision freedom," which have pertinence to locus of control. The perception of control may also be differentiated between beliefs regarding one's role in selecting a goal, and beliefs regarding the way goals are accomplished once they have been adopted. It is with this idea of greater differentiation of concepts related to locus of control that we arrive at our final point.

A number of investigators concerned with research involving perceived causation were recently convened by invitation to the National Institute of Mental Health. The purpose of the meetings that ensued was to consider plans on how it might be possible to protect this research area from suffering the usual fate of most personality constructs—to become overextended and then abandoned. The emphases of the participants varied as would be predicted. However, there was also a considerable degree of consensus about future prospects if the status quo in present research were to remain as is. If locus of control research continued at the present "general" level where currently developed scales are applied to an ever increasing range of new phenomena, it was predicted that interest in the construct would inevitably suffer a decline. Many borderline results with criteria of only marginal relevance will come to muddle the lines which separate the locus of control construct from a number of cognate concepts. With decreasing theoretical distinctions research that is germane to causal perception will have to diminish, and with it will go the stimulating value of the construct.

On the other hand, if investigators would question themselves, at the start, as to whether causal perceptions are pertinent to given criteria and purposes, and then plan upon how such perceptions are to be assessed, there would be a greater likelihood that the construct would become more finely "carved up" for particular uses. To this end, investigators have to become aware of all the finer discriminations to be made within their areas of concern. Weiner's distinctions regarding the range of sources, both internal and external, to which attributions may be made, Crandall's and Mischel's distinctions between success and failure outcome attributions, and Rotter's original distinctions with regard to reinforcement values and situational determinants that make contingency more or less likely, will all have to become essential considerations. It is here that agreement among the locus of control researchers was most evident. The construct will have its greatest utility if potential investigators design procedures for their own specific purposes and constrain themselves by formulating more appropriate and precise hypotheses with perception of control variables.

Appendices

These appendices comprise several scales that have been used in the assessment of locus of control. With each scale some normative data are presented to enable potential users to draw comparisons with their own samples. The scales are presented, however, not in the expectation that readers will come to rely upon these particular measures, but that they may be encouraged to examine these devices and the results obtained with them with a somewhat more discerning eye.

The scales presented are as follows:

1. Bialer's Locus of Control Questionnaire (Bialer, 1961);
2. Crandall's Intellectual Achievement Responsibility Questionnaire (Crandall, Katkovsky, & Crandall, 1965);
3. Dean's Alienation Scales (Dean, 1969);
4. James' I–E Scale (James, 1957);
5. Nowicki–Strickland Locus of Control Scale (Nowicki & Strickland, 1973);
6. Reid–Ware Three-Factor I–E Scale (Reid & Ware, 1974);
7. Rotter's I–E Scale (Rotter, 1966); and
8. Stanford Preschool I–E Scale (Mischel, Zeiss, & Zeiss, 1974).

In addition, I have included a new and untried device from our labs (Lefcourt, Reid, & Ware) which has been constructed to allow for a more open-ended assessment of values, expectancies, and locus of causation. This device is included because it may prove useful as a framework from which more goal-relevant locus of control measures may be derived.

APPENDIX I:
BIALER LOCUS OF CONTROL QUESTIONNAIRE*

In the administration of the questionnaire, the subject is asked to respond with "yes" or "no" to each item. The letter "f" indicates that a "yes" response is scored as internal control. The letter "p" signifies that an answer of "no" is scored as internal control. The scale is scored in terms of the total number of responses in the direction of internal control.

Children's Locus of Control Scale

Instructions

This is not a test. I am just trying to find out how kids your age think about certain things. I am going to ask you some questions to see how you feel about these things. There are no right or wrong answers to these questions. Some kids say "yes" and some say "no." When I ask the question, if you think your answer should be yes, or mostly yes, say "Yes." If you think the answer should be no, or mostly no, say "No." Remember, different children give different answers, and there is no right or wrong answer. Just say "Yes" or "No," depending on how *you* think the question should be answered. If you want me to repeat a question, ask me. Do you understand? All right, listen carefully, and answer "Yes" or "No."

1p. When somebody gets mad at you, do you usually feel there is nothing you can do about it?

2f. Do you really believe a kid can be whatever he wants to be?

3f. When people are mean to you, could it be because you did something to make them be mean?

4f. Do you usually make up your mind about something without asking someone first?

5f. Can you do anything about what is going to happen tomorrow?

6f. When people are good to you, is it usually because you did something to make them be good?

7f. Can you ever make other people do things you want them to do?

8f. Do you ever think that kids your age can change things that are happening in the world?

9f. If another child was going to hit you, could you do anything about it?

10f. Can a child your age ever have his own way?

11p. Is it hard for you to know why some people do certain things?

12f. When someone is nice to you, is it because you did the right things?

---

*From Bialer (1961).

Norms for the Bialer Locus of Control Scale

Subjects	N	Mean	SD
Normals—white, male	13	18.1	2.17
Schizophrenics—white, male	15	14.4	3.89
(Cromwell, Rosenthal, Shakow, & Zahn, 1961)			
6th and 8th grade—selected on basis			
of sex, social class, ethnic group			
from five metropolitan schools.			
Negro—middle class	16	15.8	4.4
Negro—lower class	23	18.3	3.4
White—middle class	20	15.0	4.4
White—lower class (Battle & Rotter, 1963)	21	16.4	3.5
Undergrads	108M	17.23	2.29
(A. P. MacDonald, Unpublished paper)	72F	17.41	2.12
Total:	180	17.31	2.21

13f. Can you ever try to be friends with another kid even if he doesn't want to?

14f. Does it ever help to think about what you will be when you grow up?

15f. When someone gets mad at you, can you usually do something to make him your friend again?

16f. Can kids your age ever have anything to say about where they are going to live?

17f. When you get in an argument, is it sometimes your fault?

18p. When nice things happen to you, is it only good luck?

19p. Do you often feel you get punished when you don't deserve it?

20f. Will people usually do things for you if you ask them?

21f. Do you believe a kid can usually be whatever he wants to be when he grows up?

22p. When bad things happen to you, is it usually someone else's fault?

23f. Can you ever know for sure why some people do certain things?

APPENDIX II: THE CRANDALL
INTELLECTUAL ACHIEVEMENT RESPONSIBILITY QUESTIONNAIRE*

Internal alternatives are designated by an I. Positive-events items are indicated by a plus sign, and negative events by a minus sign following the I. A child's score is obtained by summing all positive events for which he assumes credit, and his Γ score is the total of all negative events for which he assumes blame. His total I score is the sum of his I^+ and his Γ subscores.

---

*From Crandall, Katkovsky, and Crandall (1965).

The IAR Scale

1. If a teacher passes you to the next grade, would it probably be
   (a) because she liked you, or
   I^+(b) because of the work you did?

2. When you do well on a test at school, is it more likely to be
   I^+(a) because you studied for it, or
   (b) because the test was especially easy?

3. When you have trouble understanding something in school, is it usually
   (a) because the teacher didn't explain it clearly, or
   I^-(b) because you didn't listen carefully?

4. When you read a story and can't remember much of it, is it usually
   (a) because the story wasn't well written, or
   I^-(b) because you weren't interested in the story?

5. Suppose your parents say you are doing well in school. Is it likely to happen
   I^+(a) because your school work is good, or
   (b) because they are in a good mood?

6. Suppose you did better than usual in a subject at school. Would it probably happen
   I^+(a) because you tried harder, or
   (b) because someone helped you?

7. When you lose at a game of cards or checkers, does it usually happen
   (a) because the other player is good at the game, or
   I^-(b) because you don't play well?

8. Suppose a person doesn't think you are very bright or clever.
   I^-(a) Can you make him change his mind if you try to, or
   (b) are there some people who will think you're not very bright no matter what you do?

9. If you solve a puzzle quickly, is it
   (a) because it wasn't a very hard puzzle, or
   I^+(b) because you worked on it carefully?

10. If a boy or girl tells you that you are dumb, is it more likely that they say that
    (a) because they are mad at you, or
    I^-(b) because what you did really wasn't very bright?

11. Suppose you study to become a teacher, scientist, or doctor and you fail. Do you think this would happen

I⁻(a) because you didn't work hard enough, or

(b) because you needed some help and other people didn't give it to you?

12. When you learn something quickly in school, is it usually
   I⁺(a) because you paid close attention, or
   (b) because the teacher explained it clearly?

13. If a teacher says to you, "Your work is fine," is it
   (a) something teachers usually say to encourage pupils, or
   I⁺(b) because you did a good job?

14. When you find it hard to work arithmetic or math problems at school, is it
   I⁻(a) because you didn't study well enough before you tried them, or
   (b) because the teacher gave problems that were too hard?

15. When you forget something you heard in class, is it
   (a) because the teacher didn't explain it very well, or
   I⁻(b) because you didn't try very hard to remember?

16. Suppose you weren't sure about the answer to a question your teacher asked you, but your answer turned out to be right. Is it likely to happen
   (a) because she wasn't as particular as usual, or
   I⁺(b) because you gave the best answer you could think of?

17. When you read a story and remember most of it, is it usually
   I⁺(a) because you were interested in the story, or
   (b) because the story was well written?

18. If your parents tell you you're acting silly and not thinking clearly, is it more likely to be
   I⁻(a) because of something you did, or
   (b) because they happen to be feeling cranky?

19. When you don't do well on a test at school, is it
   (a) because the test was especially hard, or
   I⁻(b) because you didn't study for it?

20. When you win at a game of cards or checkers, does it happen
   I⁺(a) because you play real well, or
   (b) because the other person doesn't play well?

21. If people think you're bright or clever, is it
   (a) because they happen to like you, or
   I⁺(b) because you usually act that way?

22. If a teacher didn't pass you to the next grade, would it probably be
   (a) because she "had it in for you," or
   I⁻(b) because your school work wasn't good enough?

23. Suppose you don't do as well as usual in a subject at school. Would this probably happen
    I⁻(a) because you weren't as careful as usual, or
    (b) because somebody bothered you and kept you from working?

24. If a boy or girl tells you that you are bright, is it usually
    I⁺(a) because you thought up a good idea, or
    (b) because they like you?

25. Suppose you became a famous teacher, scientist or doctor. Do you think this would happen
    (a) because other people helped you when you needed it, or
    I⁺(b) because you worked hard?

26. Suppose your parents say you aren't doing well in your school work. Is this likely to happen more
    I⁻(a) because your work isn't very good, or
    (b) because they are feeling cranky?

27. Suppose you are showing a friend how to play a game and he has trouble with it. Would that happen
    (a) because he wasn't able to understand how to play, or
    I⁻(b) because you couldn't explain it well?

28. When you find it easy to work arithmetic or math problems at school, is it usually
    (a) because the teacher gave you especially easy problems or
    I⁺(b) because you studied your book well before you tried them?

29. When you remember something you heard in class, is it usually
    I⁺(a) because you tried hard to remember, or
    (b) because the teacher explained it well?

30. If you can't work a puzzle, is it more likely to happen
    I⁻(a) because you are not especially good at working puzzles, or
    (b) because the instructions weren't written clearly enough?

31. If your parents tell you that you are bright or clever, is it more likely
    (a) because they are feeling good, or
    I⁺(b) because of something you did?

32. Suppose you are explaining how to play a game to a friend and he learns quickly. Would that happen more often
    I⁺(a) because you explained it well, or
    (b) because he was able to understand it?

33. Suppose you're not sure about the answer to a question your teacher asks you and the answer you give turns out to be wrong. Is it likely to happen

Norms for the Crandall IAR
Questionnaire

Subjects	N	Mean	SD
Boys			
Grade 3	44	23.16	3.80
4	59	24.83	3.00
5	52	24.04	3.69
6	93	24.74	4.57
8	68	25.38	3.51
10	90	25.27	4.62
12	52	24.38	3.71
Girls			
Grade 3	58	23.22	4.00
4	44	24.75	3.81
5	47	24.36	3.96
6	73	26.93	3.71
8	93	26.64	3.86
10	93	26.50	3.93
12	57	27.33	2.98

  (a) because she was more particular than usual, or
Γ(b) because you answered too quickly?

34. If a teacher says to you, "Try to do better," would it be
  (a) because this is something she might say to get pupils to try harder, or
Γ(b) because your work wasn't as good as usual?

APPENDIX III:
DEAN'S ALIENATION SCALE(S) *

Below is a keyed copy of the Dean Alienation scale. The letter to the left of each item indicates whether it belongs to the Powerlessness, Normlessness, or Isolation subscale; scores are usually reported separately.

Public Opinion Questionnaire

Below are some statements regarding public issues, with which some people agree and others disagree. Please give us your own opinion about these items, i.e., whether you agree or disagree with the items as they stand.

---

*From Dean (1969).

Please check in the appropriate blank, as follows:

    ___A (Strongly Agree)
    ___a (Agree)
    ___U (Uncertain)
    ___d (Disagree)
    ___D (Strongly Disagree)

I   1. Sometimes I feel all alone in the world.

       _5_ A   _4_ a   _3_ U   _2_ d   _1_ D

P   2. I worry about the future facing today's children.

       _5_ A   ___a   ___U   ___d   ___D

I   3. I don't get invited out by friends as often as I'd really like.

       _5_ A   ___a   ___U   ___d   ___D

N   4. The end often justifies the means.

       _5_ A   ___a   ___U   ___d   ___D

I   5. Most people today seldom feel lonely.

       _1_ A   _2_ a   _3_ U   _4_ d   _5_ D

P   6. Sometimes I have the feeling that other people are using me.

       _5_ A   ___a   ___U   ___d   ___D

N   7. People's ideas change so much that I wonder if we'll ever have anything to depend on.

       _5_ A   ___a   ___U   ___d   ___D

I   8. Real friends are as easy as ever to find.

       _1_ A   ___a   ___U   ___d   ___D

P   9. It is frightening to be responsible for the development of a little child.

       _5_ A   ___a   ___U   ___d   ___D

N   10. Everything is relative, and there just aren't any definite rules to live by.

       _5_ A   ___a   ___U   ___d   ___D

I   11. One can always find friends if he shows himself friendly.

       _1_ A   ___a   ___U   ___d   ___D

N   12. I often wonder what the meaning of life really is.

       _5_ A   ___a   ___U   ___d   ___D

P   13. There is little or nothing I can do towards preventing a major "shoot-ing" war.

   _5_ A __a __U __d __D

I   14. The world in which we live is basically a friendly place.

   _1_ A __a __U __d __D

P   15. There are so many decisions that have to be made today that sometimes I could just "blow up."

   _5_ A __a __U __d __D

N   16. The only thing one can be sure of today is that he can be sure of nothing.

   _5_ A __a __U __d __D

I   17. There are few dependable ties between people any more.

   _5_ A __a __U __d __D

P   18. There is little chance for promotion on the job unless a man gets a break.

   _5_ A __a __U __d __D

N   19. With so many religions abroad, one doesn't really know which to believe.

   _5_ A __a __U __d __D

P   20. We're so regimented today that there's not much room for choice even in personal matters.

   _5_ A __a __U __d __D

P   21. We are just so many cogs in the machinery of life.

   _5_ A __a __U __d __D

I   22. People are just naturally friendly and helpful.

   _1_ A __a __U __d __D

P   23. The future looks very dismal.

   _5_ A __a __U __d __D

I   24. I don't get to visit friends as often as I'd really like.

   _5_ A __a __U __d __D

Mean Powerlessness Scores

	N	Scores
Columbus, Ohio (men), stratified sample, 1955.	384	22.65
Denison University, Introductory Sociology, 1962 (men)	62	23.75
Denison University, Introductory Sociology, 1965 (women)	93	24.91
Iowa State University, Social Psychology, 1971 (men)	16	24.63
Iowa State University, Social Psychology, 1971 (women)	24	24.60

APPENDIX IV: THE JAMES
INTERNAL–EXTERNAL LOCUS OF CONTROL SCALE*

The James internal–external scale is a 60-item questionnaire. The higher the score the individual obtains, the more external his orientation. (Only even numbered items are scored.)

Instructions

Below are a number of statements about various topics. They have been collected from different groups of people and represent a variety of opinions. There are no right or wrong answers to this questionnaire. For every statement there are large numbers of people who agree and disagree. Please indicate whether you agree or disagree with each statement as follows:

Circle SA	if you strongly agree	= 3
Circle A	if you agree	= 2
Circle D	if you disagree	= 1
Circle SD	if you strongly disagree	= 0

Please read each item carefully and be sure that you indicate the response which most closely corresponds to the way which you personally feel.

1. I like to read newspaper editorials whether I agree with them or not.
2. Wars between countries seem inevitable despite efforts to prevent them.
3. I believe the government should encourage more young people to make science a career.
4. It is usually true of successful people that their good breaks far outweigh their bad breaks.

*From James (1957).

5. I believe that moderation in all things is the key to happiness.
6. Many times I feel that we might just as well make many of our decisions by flipping a coin.
7. I disapprove of girls who smoke cigarettes in public places.
8. The actions of other people toward me many times have me baffled.
9. I believe it is more important for a person to like his work than to make money at it.
10. Getting a good job seems to be largely a matter of being lucky enough to be in the right place at the right time.
11. It's not what you know but who you know that really counts in getting ahead.
12. A great deal that happens to me is probably just a matter of chance.
13. I think that people spend too much time watching television these days.
14. I feel that I have little influence over the way people behave.
15. It is difficult for me to keep well-informed about foreign affairs.
16. Much of the time the future seems uncertain to me.
17. I think the world is much more unsettled now than it was in our grand-father's times.
18. Some people seem born to fail while others seem born for success no matter what they do.
19. I believe there should be less emphasis on spectator sports and more on athletic participation.
20. It is difficult for ordinary people to have much control over what politicians do in office.
21. I tend to daydream more than I should.
22. I feel that many people could be described as victims of circumstances beyond their control.
23. Movies do not seem as good as they used to be.
24. It seems many times that the grades one gets in school are more dependent on the teacher's whims than on what the student can really do.
25. Money shouldn't be a person's main consideration in choosing a job.
26. It isn't wise to plan too far ahead because most things turn out to be a matter of good or bad fortune anyhow.
27. At one time I wanted to become a newspaper reporter.
28. I can't understand how it is possible to predict other people's behaviour.
29. I enjoy smoking cigarettes and will continue to be a smoker.
30. When things are going well for me I consider it due to a run of good luck.
31. I believe the federal government has been taking over too many of the affairs of private management.
32. There's not much use in trying to predict which questions a teacher is going to ask on an examination.
33. I get more ideas from talking about things than reading about them.
34. Most people don't realize the extent to which their lives are controlled by accidental happenings.

Norms for the James Internal–External Scale

Subjects	N	Mean	SD
Normals	13	51.7	7.66
Schizophrenics	15	42.3	4.98
(Cromwell, Rosenthal, Shakow, & Zahn, 1961)			
Undergrads	33F	41.6	8.8
(Rotter, Simmons, & Holden, 1962)	31M	37.68	8.1
Undergrads—half male and half female	80	37	12
(Lotsof, James, Drucker, & Blount, 1964)			
Physically Disabled	30	41.6	3.3
Non Disabled	30	40.3	3.57
(Lipp, Kolstoe, James, & Randall, 1968)			
Undergrads	108M	43.09	9.24
(A. P. MacDonald, unpublished)	72F	44.19	11.18

35. At one time I wanted to be an actor (or actress).
36. I have usually found that what is going to happen will happen, regardless of my actions.
37. Life in a small town offers more real satisfactions than life in a large city.
38. Most of the disappointing things in my life have contained a large element of chance.
39. I would rather be a successful teacher than a successful businessman.
40. I don't believe that a person can really be a master of his fate.
41. I find mathematics easier to study than literature.
42. Success is mostly a matter of getting good breaks.
43. I think it is more important to be respected by people than to be liked by them.
44. Events in the world seem to be beyond the control of most people.
45. I think our country should take a more active role in world affairs.
46. I feel that most people can't really be held responsible for themselves since no one has much choice about where he was born or raised.
47. I like to figure out problems and puzzles that other people have trouble with.
48. Many times the reactions of people seem haphazard to me.
49. I rarely lose when playing card games.
50. There's not much use in worrying about things—what will be, will be.
51. I think that everyone should belong to some kind of church.
52. Success in dealing with people seems to be more a matter of the other person's moods and feelings at the time rather than one's own actions.
53. One should not place too much faith in newspaper reports.
54. I think that life is mostly a gamble.

55. I am very stubborn when my mind is made up about something.
56. Many times I feel that I have little influence over the things that happen to me.
57. I like popular music better than classical music.
58. Sometimes I feel that I don't have enough control over the direction my life is taking.
59. I sometimes work at difficult things too long even when I know they are hopeless.
60. Life is too full of uncertainties.

APPENDIX V: THE NOWICKI–STRICKLAND
LOCUS OF CONTROL SCALE*

The Nowicki–Strickland Locus of Control scale is a paper and pencil measure consisting of 40 questions which are answered either yes or no by placing a mark next to the question. The higher the score the more external the orientation.

Nowicki–Strickland Scale

(Y) 1. Do you believe that most problems will solve themselves if you just don't fool with them?

(N) 2. Do you believe that you can stop yourself from catching a cold?

(Y) 3. Are some kids just born lucky?

(N) 4. Most of the time do you feel that getting good grades means a great deal to you?

(Y) 5. Are you often blamed for things that just aren't your fault?

(N) 6. Do you believe that if somebody studies hard enough he or she can pass any subject?

(Y) 7. Do you feel that most of the time it doesn't pay to try hard because things never turn out right anyway?

(Y) 8. Do you feel that if things start out well in the morning that it's going to be a good day no matter what you do?

(N) 9. Do you feel that most of the time parents listen to what their children have to say?

(Y) 10. Do you believe that wishing can make good things happen?

(Y) 11. When you get punished does it usually seem it's for no good reason at all?

(Y) 12. Most of the time do you find it hard to change a friend's (mind) opinion?

(N) 13. Do you think that cheering more than luck helps a team to win?

---

*From Nowicki and Strickland (1973).

(Y) 14. Do you feel that it's nearly impossible to change your parent's mind about anything?

(N) 15. Do you believe that your parents should allow you to make most of your own decisions?

(Y) 16. Do you feel that when you do something wrong there's very little you can do to make it right?

(Y) 17. Do you believe that most kids are just born good at sports?

(Y) 18. Are most of the other kids your age stronger than you are?

(Y) 19. Do you feel that one of the best ways to handle most problems is just not to think about them?

(N) 20. Do you feel that you have a lot of choice in deciding who your friends are?

(Y) 21. If you find a four-leaf clover do you believe that it might bring you good luck?

(N) 22. Do you often feel that whether you do your homework has much to do with what kind of grades you get?

(Y) 23. Do you feel that when a kid your age decides to hit you, there's little you can do to stop him or her?

(Y) 24. Have you ever had a good luck charm?

(N) 25. Do you believe that whether or not people like you depends on how you act?

(N) 26. Will your parents usually help you if you ask them to?

(Y) 27. Have you felt that when people were mean to you it was usually for no reason at all?

(N) 28. Most of the time, do you feel that you can change what might happen tomorrow by what you do today?

(Y) 29. Do you believe that when bad things are going to happen they just are going to happen no matter what you try to do to stop them?

(N) 30. Do you think that kids can get their own way if they just keep trying?

(Y) 31. Most of the time do you find it useless to try to get your own way at home?

(N) 32. Do you feel that when good things happen they happen because of hard work?

(Y) 33. Do you feel that when somebody your age wants to be your enemy there's little you can do to change matters?

(N) 34. Do you feel that it's easy to get friends to do what you want them to?

(Y) 35. Do you usually feel that you have little to say about what you get to eat at home?

(Y) 36. Do you feel that when someone doesn't like you there's little you can do about it?

(Y) 37. Do you usually feel that it's almost useless to try in school because most other children are just plain smarter than you are?

(N) 38. Are you the kind of person who believes that planning ahead makes things turn out better?

The Nowicki–Strickland
Locus of Control Scale

Subjects	N	Mean	SD
Males			
grade 3	44	17.97	4.67
grade 4	59	18.44	3.58
grade 5	40	18.32	4.38
grade 6	45	13.73	5.16
grade 7	65	13.15	4.87
grade 8	75	14.73	4.35
grade 9	43	13.81	4.06
grade 10	68	13.05	5.34
grade 11	37	12.48	4.81
grade 12	39	11.38	4.74
Females			
grade 3	55	17.38	3.06
grade 4	45	18.80	3.63
grade 5	41	17.00	4.03
grade 6	43	13.32	4.58
grade 7	52	13.94	4.23
grade 8	34	12.29	3.58
grade 9	44	12.25	3.75
grade 10	57	12.98	5.31
grade 11	53	12.01	5.15
grade 12	48	12.37	5.05

(Y) 39. Most of the time, do you feel that you have little to say about what your family decides to do?

(N) 40. Do you think it's better to be smart than to be lucky?

APPENDIX VI: THE REID–WARE
THREE-FACTOR INTERNAL–EXTERNAL SCALE*

The Reid–Ware scale is a 45-item forced choice questionnaire composed of the following three factors: self-control, social systems control, and fatalism. The higher the score the more external the individual. There are 13 filler items.

Belief Survey

This questionnaire is a measure of personal belief: obviously there are no right or wrong answers. Each item consists of a pair of alternatives lettered (A) or (B). Please select the one statement of each pair (and only one) which you more

---

*From Reid and Ware (1974).

strongly believe as far as you are concerned. Be sure to select the one you actually believe to be more true rather than the one you think you should choose or the one you would like to be true. To ensure anonymity, please use a fictitious name when signing this questionnaire.

Please answer these items carefully, but do not spend too much time on any one item. Be sure to find an answer for every choice. Circle the letter of the statement (A or B) which you choose.

In some cases you may discover that you believe both statements or neither one. In such cases be sure to select the one you more strongly believe to be the case as far as you are concerned. Also try to respond to each item independently when making your choice: do not be influenced by your previous choices.

1. (A) Various sports activities in the community help increase solidarity amongst people in the community.
   (B) Various sports activities in the community can lead to rivalry detrimental to the solidarity of the community.

2. (A) War brings out the worst aspects of men.
   (B) Although war is terrible, it can have some value.

SSC E   3. (A) There will always be wars no matter how hard people try to prevent them.
   (B) One of the major reasons why we have wars is because people do not take enough interest in politics.

SC E   4. (A) Even when there was nothing forcing me, I have found that I will sometimes do things I really did not want to do.
   (B) I always feel in control of what I am doing.

SSC E   5. (A) There are institutions in our society that have considerable control over me.
   (B) Little in this world controls me, I usually can do what I decide to do.

6. (A) I would like to live in a small town or a rural environment.
   (B) I would like to live in a large city.

Fatalism   7. (A) For the average citizen becoming a success is a matter of hard work, luck has little or nothing to do with it.
   E   (B) For the average guy getting a good job depends mainly on being in the right place at the right time.

8. (A) Patriotism demands that the citizens of a nation participate in any war.
   (B) To be a patriot for one's country does not necessarily mean he must go to war for his country.

Fatalism   9. (A) In my case getting what I want has little or nothing to do with luck.

E   (B) It is not always wise for me to plan too far ahead because many things turn out to be a matter of good or bad fortune anyhow.

SC E 10. (A) Sometimes I impulsively do things which at other times I definitely would not let myself do.

(B) I find that I can keep my impulses in control.

Fatalism E 11. (A) In many situations what happens to people seems to be determined by fate.

(B) People do not realize how much they personally determine their own outcomes.

12. (A) College students should be trained in times of peace to assume military duties.

(B) The ills of war are greater than any possible benefits.

Fatalism E 13. (A) Most people do not realize the extent to which their lives are controlled by accidental happenings.

(B) For any guy, there is no such thing as luck.

SSC   14. (A) If I put my mind to it I could have an important influence on what a politician does in office.

E   (B) When I look at it carefully I realize it is impossible for me to have any really important influence over what politicians do.

Fatalism E 15. (A) With fate the way it is, many times I feel that I have little influence over the things that happen to me.

(B) It is impossible for me to believe that chance or luck plays an important role in my life.

SC   16. (A) When I put my mind to it I can constrain my emotions.

E   (B) There are moments when I cannot subdue my emotions and keep them in check.

17. (A) Every person should give some of his time for the good of his town or country.

(B) People would be a lot better off if they could live far away from other people and never have to do anything for them.

SSC E 18. (A) As far as the affairs of our country are concerned, most people are the victims of forces they do not control and frequently do not even understand.

           (B) By taking part in political and social events the people can directly control much of the country's affairs.

SC E 19. (A) People cannot always hold back their personal desires; they will behave out of impulse.

           (B) If they want to, people can always control their immediate wishes and not let these motives determine their total behavior.

Fatalism E 20. (A) Many times I feel I might just as well decide what to do by flipping a coin.

           (B) In most cases I do not depend on luck when I decide to do something.

        21. (A) Our federal government should promote the mass production of low rental apartment buildings to reduce the housing shortage.

           (B) The best way for our government to reduce the housing shortage is to make low interest mortgages available and to stimulate the building of low cost houses.

SSC E 22. (A) I do not know why politicians make the decisions they do.

           (B) It is easy for me to understand why politicians do the things they do.

SC 23. (A) Although sometimes it is difficult, I can always willfully restrain my immediate behavior.

    E (B) Something I cannot do is have complete mastery over all my behavioral tendencies.

Fatalism 24. (A) In the long run people receive the respect and good outcomes they worked for.

    E (B) Unfortunately, because of misfortune or bad luck, the average guy's worth often passes unrecognized no matter how hard he tries.

SSC 25. (A) With enough effort people can wipe out political corruption.

    E (B) It is difficult for people to have much control over the things politicians do in office.

        26. (A) Letting your friends down is not so bad because you cannot do good all the time for everybody.

           (B) I feel very bad when I have failed to finish a job I promised I would do.

SSC   27. (A) By active participation in the appropriate political organizations people can do a lot to keep the cost of living from
going higher.

E   (B) There is very little people can do to keep the cost of living
from going higher.

SC E   28. (A) It is possible for me to behave in a manner very different
from the way I would want to behave.

(B) It would be very difficult for me to not have mastery over
the way I behave.

SSC E 29. (A) In this world I am affected by social forces which I
neither control nor understand.

(B) It is easy for me to avoid and function independently of
any social forces that may attempt to have control over
me.

30. (A) It hurts more to lose money than to lose a friend.

(B) The people are the most important thing in this world of
ours.

Fatalism   31. (A) What people get out of life is always a function of how
much effort they put into it.

E   (B) Quite often one finds that what happens to people has no
relation to what they do, what happens just happens.

SSC   32. (A) Generally speaking, my behavior is not governed by
others.

E   (B) My behavior is frequently determined by other influential
people.

SSC   33. (A) People can and should do what they want to do both now
and in the future.

E   (B) There is no point in people planning their lives too far in
advance because other groups of people in our society will
invariably upset their plans.

34. (A) Happiness is having your own house and car.

(B) Happiness to most people is having their own close
friends.

Fatalism   35. (A) There is no such thing as luck, what happens to me is a
result of my own behavior.

E   (B) Sometimes I do not understand how I can have such poor
luck.

36. (A) More emphasis should be placed on teaching the principles of Christianity in public school.
    (B) Christianity should not be included in a school curriculum; it can be taught in church.

Fatalism E 37. (A) Many of the unhappy things in people's lives are at least partly due to bad luck.
    (B) People's misfortunes result from the mistakes they make.

SC 38. (A) Self-regulation of one's behavior is always possible.
E   (B) I frequently find that when certain things happen to me I cannot restrain my reaction.

SSC 39. (A) The average man can have an influence in government decisions.
E   (B) This world is run by a few people in power and there is not much the little guy can do about it.

SC 40. (A) When I make up my mind, I can always resist temptation and keep control of my behavior.
E   (B) Even if I try not to submit, I often find I cannot control myself from some of the enticements in life such as over-eating or drinking.

Fatalism E 41. (A) My getting a good job or promotion in the future will depend a lot on my getting the right turn of fate.
    (B) When I get a good job, it is always a direct result of my own ability and/or motivation.

42. (A) Successful people are mostly honest and good.
    (B) One should not always associate achievement with integrity and honor.

SSC E 43. (A) Most people do not understand why politicians behave the way they do.
    (B) In the long run people are responsible for bad government on a national as well as on a local level.

Fatalism E 44. (A) I often realize that despite my best efforts some outcomes seem to happen as if fate planned it that way.
    (B) The misfortunes and successes I have had were the direct result of my own behavior.

45. (A) Most people are kind and good.
    (B) People will not help others unless circumstances force them to.

Norms for the Reid–Ware Three-Factor
Internal–External Scale

Subjects	N	Mean	SD
Undergrads (Reid & Ware, 1973)			
Males			
Self-control factor	55	4.25	2.29
Social systems control		6.65	2.98
Fatalism factor		4.09	2.88
Females			
Self-control factor	44	5.52	1.91
Social systems control		6.84	2.75
Fatalism		4.47	2.88
Undergrads: Male and Female (Reid & Ware, 1973)	167		
Self-control factor		4.77	2.25
Social systems control		6.53	2.82
Fatalism factor		3.99	2.86

APPENDIX VII: THE ROTTER INTERNAL–EXTERNAL LOCUS OF CONTROL*

The Rotter internal–external locus of control scale is a 23-item forced choice questionnaire with 6 filler items adapted from the 60-item James scale. It is scored in the external direction, that is, the higher the score the more external the individual.

Social Reaction Inventory

This is a questionnaire to find out the way in which certain important events in our society affect different people. Each item consists of a pair of alternatives lettered a or b. Please select the one statement of each pair (and only one) which you more strongly *believe* to be the case as far as you're concerned. Be sure to select the one you actually believe to be more true rather than the one you think you should choose or the one you would like to be true. This is a measure of personal belief; obviously there are no right or wrong answers.

Your answer, either a or b to each question on this inventory, is to be reported beside the question. Print your name and any other information requested by the examiner on the bottom of page 4, then finish reading these directions. Do not begin until you are told to do so.

*From Rotter (1966).

Please answer these items *carefully* but do not spend too much time on any one item. Be sure to find an answer for *every* choice. For each numbered question make an X on the line beside either the *a* or *b*, whichever you choose as the statement most true.

In some instances you may discover that you believe both statements or neither one. In such cases, be sure to select the one you more strongly believe to be the case as far as you're concerned. Also try to respond to each item *independently* when making your choice; do not be influenced by your previous choices.

Remember

Select that alternative which you *personally believe to be more true.*

I more strongly believe that:

1. __ a. Children get into trouble because their parents punish them too much.
   __ b. The trouble with most children nowadays is that their parents are too easy with them.

E 2. __ a. Many of the unhappy things in people's lives are partly due to bad luck.
   __ b. People's misfortunes result from the mistakes they make.

3. __ a. One of the major reasons why we have wars is because people don't take enough interest in politics.
E __ b. There will always be wars, no matter how hard people try to prevent them.

4. __ a. In the long run people get the respect they deserve in this world.
E __ b. Unfortunately, an individual's worth often passes unrecognized no matter how hard he tries.

5. __ a. The idea that teachers are unfair to students is nonsense.
E __ b. Most students don't realize the extent to which their grades are influenced by accidental happenings.

E 6. __ a. Without the right breaks one cannot be an effective leader.
   __ b. Capable people who fail to become leaders have not taken advantage of their opportunities.

E 7. __ a. No matter how hard you try some people just don't like you.
   __ b. People who can't get others to like them don't understand how to get along with others.

8. ___ a. Heredity plays the major role in determining one's personality.
   ___ b. It is one's experiences in life which determine what they're like.

E 9. ___ a. I have often found that what is going to happen will happen.
   ___ b. Trusting to fate has never turned out as well for me as making a decision to take a definite course of action.

10. ___ a. In the case of the well prepared student there is rarely if ever such a thing as an unfair test.

E ___ b. Many times exam questions tend to be so unrelated to course work that studying is really useless.

11. ___ a. Becoming a success is a matter of hard work, luck has little or nothing to do with it.

E ___ b. Getting a good job depends mainly on being in the right place at the right time.

12. ___ a. The average citizen can have an influence in government decisions.

E ___ b. This world is run by the few people in power, and there is not much the little guy can do about it.

13. ___ a. When I make plans, I am almost certain that I can make them work.

E ___ b. It is not always wise to plan too far ahead because many things turn out to be a matter of good or bad fortune anyhow.

14. ___ a. There are certain people who are just no good.
   ___ b. There is some good in everybody.

15. ___ a. In my case getting what I want has little or nothing to do with luck.

E ___ b. Many times we might just as well decide what to do by flipping a coin.

E 16. ___ a. Who gets to be the boss often depends on who was lucky enough to be in the right place first.
   ___ b. Getting people to do the right thing depends upon ability; luck has little or nothing to do with it.

E 17. ___ a. As far as world affairs are concerned, most of us are the victims of forces we can neither understand, nor control.
   ___ b. By taking an active part in political and social affairs the people can control world events.

E 18. ___ a. Most people can't realize the extent to which their lives are controlled by accidental happenings.

          __ b. There really is no such thing as "luck."

19. __ a. One should always be willing to admit his mistakes.
      __ b. It is usually best to cover up one's mistakes.

E 20. __ a. It is hard to know whether or not a person really likes you.
        __ b. How many friends you have depends upon how nice a person you are.

E 21. __ a. In the long run the bad things that happen to us are balanced by the good ones.
        __ b. Most misfortunes are the result of lack of ability, ignorance, laziness, or all three.

22. __ a. With enough effort we can wipe out political corruption.
E   __ b. It is difficult for people to have much control over the things politicians do in office.

E 23. __ a. Sometimes I can't understand how teachers arrive at the grades they give.
        __ b. There is a direct connection between how hard I study and the grades I get.

24. __ a. A good leader expects people to decide for themselves what they should do.
      __ b. A good leader makes it clear to everybody what their jobs are.

E 25. __ a. Many times I feel that I have little influence over the things that happen to me.
        __ b. It is impossible for me to believe that chance or luck plays an important role in my life.

26. __ a. People are lonely because they don't try to be friendly.
E   __ b. There's not much use in trying too hard to please people, if they like you, they like you.

27. __ a. There is too much emphasis on athletics in high school.
      __ b. Team sports are an excellent way to build character.

28. __ a. What happens to me is my own doing.
E   __ b. Sometimes I feel that I don't have enough control over the direction my life is taking.

E 29. __ a. Most of the time I can't understand why politicians behave the way they do.
        __ b. In the long run the people are responsible for bad government on a national as well as on a local level.

Norms for the Rotter Internal–External
Locus of Control Scale

Scores are in the external direction, the higher the score the more external.

Subjects	N	Mean	SD
Students at a Southern Negro college	62M		
involved in protest movements	54F		
(Gore & Rotter, 1963)			
1. Attend rally for civil rights		10.3	3.1
2. Sign petition		9.2	3.4
3. Join a silent march		7.4	2.9
4. Join Freedom Riders		8.1	3.8
5. None of the above		10.0	3.9
Inmates of correctional institution			
(Lefcourt & Ladwig, 1965a)			
1. Negro	60	8.97	2.97
2. White	60	7.87	3.03
Negro college students—male and female			
(Strickland, 1965)			
1. Active—engaged in civil rights			
groups	53	7.49	3.49
2. Inactive	105	9.64	3.70
1. 1964 Service Corps	72F	7.92	3.84
	27M	8.00	3.97
2. 1965 Service Corps	68F	8.26	3.49
	34M	8.00	3.08
3. 1965 Control Group	46F	9.37	3.76
	49M	8.67	3.89
4. 1966 Service Corps	79F	9.54	4.20
	21M	7.38	4.73
5. 1966 Control Group	47F	8.79	3.76
(Hersch & Scheibe, 1967)			
	38M	8.84	3.70

(Service Corps were college students attending chronic wards of mental institutions.) (Control Groups were college students attending summer school.)

Undergrads (Levy, 1967):	24M		
	& 24F	9.77	4.11
Male & female smokers—mean age	213	7.0	3.50
of 40.1; average of 13.4 years of	95M	6.59	3.65
education (Lichtenstein & Keutzer, 1967)	118F	7.42	3.44
College males (Zytowski, 1967)	62	6.82	2.49
Undergrads in introductory psychology	46M	9.8	1.42
(Feather, 1968)	88F	11.44	1.69

continued

Norms for the Rotter Scale — *Continued*

Subjects	*N*	Mean	*SD*
Undergrads (Hamsher, Geller, & Rotter, 1968)	60M	10.1	3.95
	113F	11.0	3.96
Undergrads—male and female; males made up 70% of sample with no significant sex differences (Julian & Katz, 1968)	1338	8.4	4.12
High school students (Hsieh, Shybut & Lotsof, 1969)			
1. Anglo-American	131M	8.58	3.89
	108F		
2. American-born Chinese	38M	9.79	3.07
	42F		
3. Hong Kong students	241M	12.07	3.96
	102F		
Male addict patients—Negro and white (Berzins & Ross, 1973)	97	6.79	3.90
Female undergrads (Crego, 1970)	99	7.97	3.8
First year female undergrads unable to relate in interpersonal situations (Dua, 1970)			
1. Pretest	30	14.03	4.27
2. Posttest	30	9.66	3.59
Female student nurses (Lefcourt & Steffy, 1970)	37	7.14	3.28
Female undergrads (Strickland, 1970)	180	8.34	3.85
Undergrads enrolled in introductory psychology, male and female (Biondo & MacDonald, 1971)	198	9.56	
Male soldiers (Cone, 1971)			
1. Mental clinic outpatients—all male soldiers	102	12.64	8.33
2. Stockade prisoners—soldiers	110	12.20	7.84
3. Same as 2 but tested 2 months later	98	12.87	7.76
Administrators (Harvey, 1971)			
1–5 years	14	7.57	2.88
6–10 years	7	6.43	2.52
11 years	27	5.41	3.15
1–10 years	21	7.19	2.75
Male VA psychiatric patients (Kish, Solberg, & Uecker, 1971)	169	8.1	4.2
Male undergrads (Lefcourt & Telegdi, 1971)	90	8.16	4.38

continued

Norms for the Rotter Scale — *Continued*

Subjects	N	Mean	SD
Hospitalized male veterans (Palmer, 1971)			
1. Psychiatric	89	5.0	2.77
2. Nonpsychiatric	88	4.0	2.70
Males in introductory psychology classes (Phares, 1971)	646	9.2	3.48
Undergrads in psychology or social science classes (Schneider & Parsons, 1970)			
Males:			
United States	95	9.76	
West Germans	44	9.75	
Denmark	124	9.83	
Japan	67	13.45	
Females:			
United States	74	10.38	
West Germans	24	10.96	
Denmark	147	9.94	
Japan	41	14.40	

APPENDIX VIII: THE STANFORD PRESCHOOL INTERNAL–EXTERNAL SCALE*

The SPIES measures preschool children's expectancies about whether events occur as a consequence of their own action (internal control) or as a consequence of external forces (external control). A forced choice format elicits expectancies about locus of control separately for positive and negative events. The SPIES is scored in the internal direction; scores obtained are expectancies for internal control of positive events (I^+) and negative events (I^-) and a sum of these two (total I).

The Stanford Preschool Internal–External Scale (I)

1. When you are happy, are you happy
   I^+(a) because you did something fun, or
   (b) because somebody was nice to you?

2. When somebody tells you that you are good, is that
   I^+(a) because you really have been good, or
   (b) because he is a nice person?

---

*From Mischel, Zeiss, and Zeiss (1974).

3. Do you think I brought you to the surprise room (experimental room)
   I^+(a) because you have been good today, or
   (b) because I'm just a nice guy (lady)?

4. When your mother gives you a cookie, is that
   I^+(a) because you need a cookie, or
   (b) because she has too many cookies?

5. When somebody brings you a present, is that
   I^+(a) because you are a good boy (girl), or
   (b) because they like to give people presents?

6. When you draw a whole picture without breaking your crayon is that
   I^+(a) because you were very careful, or
   (b) because it was a good crayon?

7. If you had a shiny new penny and lost it, would that be
   I^-(a) because you dropped it, or
   (b) because there was a hole in your pocket?

8. When you are sad and unhappy, are you sad and unhappy
   I^-(a) because you did something sad, or
   (b) because somebody wasn't very nice to you?

9. When you play a game and lose, do you lose
   I^-(a) because you just didn't play well, or
   (b) because the game was hard?

10. When somebody stops playing with you, is that
    I^-(a) because he doesn't like the way you play, or
    (b) because he is tired?

11. When you get a hole in your pants, is that
    I^-(a) because you tore them or
    (b) because they wore out?

12. If you had a pet turtle and he ran away, do you think that would be
    I^-(a) because you did something to make him leave, or
    (b) because there was a hole in his cage?

13. When you are drawing a picture and your crayon breaks, is that
    I^-(a) because you pushed too hard, or
    (b) because it was a bad crayon?

14. When you can't find one of your toys, is that
    I^-(a) because you lost it, or
    (b) because somebody took it?

The Stanford Preschool Internal–External Scale (II)

1. When you're playing at a friend's house and his mother gives you a piece of candy, is that
   I⁺(a) because you want some candy, or
   (b) because she wants to be nice?

2. When your father reads a book to you, is that
   I⁺(a) because you want to hear the story, or
   (b) because he likes to read?

3. If you put a hard puzzle together, is that
   I⁺(a) because you knew how, or
   (b) because your teacher helped you?

4. If you walk into the room and your mother smiles, is that
   I⁺(a) because you did something to make her smile, or
   (b) because she likes to smile a lot?

5. If you wake up from a nap and your mother lets you get up, is that
   I⁺(a) because you slept enough, or
   (b) because she thinks it's time to get up?

6. If you're good and eat all your dinner, are you good
   I⁺(a) because you're hungry, or
   (b) because the food is good?

7. If you pet a dog and he's happy and wags his tail, is that
   I⁺(a) because you were friendly and nice to him, or
   (b) because he likes all children?

8. If you pet a dog and he runs away, is that
   I⁻(a) because you scared him or
   (b) because he doesn't like any children?

9. If you're playing and your friend throws you a ball and it goes out into the street, is that
   I⁻(a) because you missed it, or
   (b) because he threw it too hard?

10. When you fall down and hurt yourself, is that
    I⁻(a) because you weren't careful, or
    (b) because somebody was in your way?

11. If you wake up early from a nap and your mother tells you to stay in bed longer, did you wake up
    I⁻(a) because you weren't trying hard enough to sleep, or
    (b) because your mother made too much noise?

Norms for the SPIES
(Mischel, Zeiss, & Zeiss, 1974)

Subjects	N	Mean	SD
Males:	98		
1. I^+		2.24	1.20
2. I^-		2.55	1.41
3. Total I		5.80	1.88
Females	113		
1. I^+		3.31	1.19
2. I^-		2.71	1.52
3. Total I		6.02	1.87
Combined	211		
1. I^+		3.28	1.19
2. I^-		2.64	1.47
3. Total I		5.91	1.87

12. If you're bad and play with your food and don't eat your dinner, are you bad
    I^-(a) because you're not hungry, or
    (b) because the food is not good?

13. If you had a toy that was broken and didn't work, would that be
    I^-(a) because you dropped it, or
    (b) because it wore out?

14. When your mother scolds you, is that
    I^-(a) because you did something wrong, or
    (b) because she was just mad?

APPENDIX IX: LEFCOURT, REID, AND WARE
INTERVIEW QUESTIONS FOR ASSESSING VALUES, EXPECTANCIES, AND LOCUS OF CONTROL

This newly developed questionnaire has been designed to be administered either as an interview or as an extensive, open-ended questionnaire. It has the benefit of allowing for causal attributions to success and failure for valued goals. It is hoped that this device will encourage others to invent more goal-specific methods such as these for their own purposes.

1. For one, what about your present life is of the greatest importance to you? Or, to put it another way, what essentials are there in your present life which you would find most distressing to lose?

2. What about your life would change if this (these) loss (losses) were to occur? That is, what would this (these) loss (losses) mean to you?
3. Of the essentials in your life that you have described, which is *the* most important for you?
4. What chance do you feel there is of losing this "essential" in the near future? (Check one point.)

___	___	___	___	___
highly unlikely	unlikely	possibly	likely	highly likely

5. What would contribute to the successful continuation of this major concern of yours?
6. If you were to lose out with regard to this "essential" what would be the most likely causes of this loss?
7. What things do you most look forward to in your own future?
8. Which of these ambitions is of most concern to you?
9. What's important about it for you?
10. How likely is it that your hopes will be met? (Check one point.)

___	___	___	___	___
very unlikely	unlikely	possibly	likely	very likely

11. If you were to succeed in your ambitions, to what would you attribute your success? What would be the most prominent of these causes?
12. If you were to fail to realize your ambitions, what would be the likely causes? Which would be the most important cause?
13. How important would it be to you to know that you were largely responsible for attaining success with your major concerns? Why?
14. And, how important would it be for you to know that you were largely responsible for failure with regard to your major concerns? Why?

References

Abramowitz, S. I. Locus of control and self-reported depression among college students. *Psychological Reports,* 1969, **25,** 149–150.

Abramowitz, S. I. Internal–external control and social political activism: A test of the dimensionality of Rotter's internal–external scale. *Journal of Consulting and Clinical Psychology,* 1973, **40,** 196–201.

Alpert, R., & Haber, R. N. Anxiety in academic achievement situations. *Journal of Abnormal and Social Psychology,* 1960, **61,** 207–215.

Ansbacher, H. & Ansbacher, R. *The individual psychology of Alfred Adler.* New York: Basic Books, 1956.

Antrobus, P. M. Internality, pride and humility. Unpublished doctoral dissertation, University of Waterloo, 1973.

Arendt, H. *Adolph Eichmann in Jerusalem: A report on the banality of evil.* New York: Viking Press, 1963.

Averill, J. R. Personal control over aversive stimuli and its relationship to stress. *Psychological Bulletin,* 1973, **80,** 286–303.

Baron, R. M., Cowan, G., Ganz, R. L., & McDonald, M. Interaction of locus of control and type of performance feedback: Considerations of external validity. *Journal of Personality and Social Psychology,* 1974, **30,** 285–292.

Baron, R. M., & Ganz, R. L. Effects of locus of control and type of feedback on the task performance of lower class black children. *Journal of Personality and Social Psychology,* 1972, **21,** 124–130.

Battle, E., & Rotter, J. B. Children's feelings of personal control as related to social class and ethnic groups. *Journal of Personality,* 1963, **31,** 482–490.

Bax, J. C. Internal–external control and field dependence. Unpublished honors thesis, University of Waterloo, 1966.

Beck, A. T. *Depression: Causes and treatment.* Philadelphia: University of Pennsylvania Press, 1967.

Berzins, J. I., & Ross, W. F. Locus of control among opiate addicts. *Journal of Consulting and Clinical Psychology,* 1973, **40,** 84–91.

Bettelheim, B. Individual and mass behavior in extreme situations. *Journal of Abnormal and Social Psychology,* 1943, **38,** 417–452.

Bialer, I. Conceptualization of success and failure in mentally retarded and normal children. *Journal of Personality,* 1961, **29,** 303–320.

Bibring, E. The mechanism of depression. In P. Greenacre (Ed.), *Affective disorders.* New York: International Universities Press, 1953.

Biondo, J., & MacDonald, A. P. Internal–external locus of control and response to influence attempts. *Journal of Personality,* 1971, **39,** 407–419.

Bowers, K. S. Pain, anxiety, and perceived control. *Journal of Consulting and Clinical Psychology,* 1968, **32,** 596–602.

Brady, J. V., Porter, R. W., Conrad, D. G., & Mason, J. W. Avoidance behavior and the development of gastroduodenal ulcers. *Journal of Experimental Analyses of Behavior,* 1958, **1,** 69–72.

Brecher, M., & Denmark, F. L. Internal–external locus of control and verbal fluency. *Psychological Reports,* 1969, **25,** 707–710.

Burnes, K., Brown, W. A., & Keating, G. W. Dimensions of control: Correlations between MMPI and I–E scores. *Journal of Consulting and Clinical Psychology,* 1971, **36,** 301.

Buss, A. H. *Psychopathology.* New York: Wiley, 1966.

Butterfield, E. C. Locus of control, test anxiety, reactions to frustration, and achievement attitudes. *Journal of Personality,* 1964, **32,** 298–311.

Butterfield, E. C. The role of competence motivation in interrupted task recall and repetition choice. *Journal of Experimental Child Psychology,* 1965, **2,** 354–370.

Chance, J. E. Internal control of reinforcements and the school learning process. Paper presented at Society for Research in Child Development Convention, Minneapolis, 1965.

Chance, J. E., & Goldstein, A. G. Locus of control and performance on embedded figures. Paper presented at the Midwestern Psychological Association Convention, Chicago, 1967. Also in *Perception and Psychophysics,* 1971, **9,** 33–34.

Coleman, J. S. *Resources for social change: Race in the United States.* Toronto: Wiley (Interscience), 1971.

Coleman, J. S., Campbell, E. Q., Hobson, C. J., McPartland, J., Mood, A. M., Weinfeld, F. D., & York, R. L. *Equality of educational opportunity.* Washington, D.C.: United States Government Printing Office, 1966.

Collins, B. E. Four separate components of the Rotter I–E Scale: Belief in a difficult world, a just world, a predictable world, and a politically responsive world. *Journal of Personality and Social Psychology,* 1974, **29,** 381–391.

Collins, B. E., Martin, J. C., Ashmore, R. D., & Ross, L. Some dimensions of the internal–external metaphor in theories of personality, *Journal of Personality,* 1973, **41,** 471–492.

Cone, J. D. Locus of control and social desirability, *Journal of Consulting and Clinical Psychology,* 1971, **36,** 449.

Cooper, J. F. *The deerslayer.* Boston: Ginn & Co., 1910.

Corah, J. L., & Boffa, J. Perceived control, self-observation and response to aversive stimulation. *Journal of Personality and Social Psychology,* 1970, **16,** 1–14.

Crandall, V. C., Katkovsky, W., & Crandall, V. J. Children's beliefs in their control of reinforcements in intellectual academic achievement behaviors. *Child Development,* 1965, **36,** 91–109.

Crandall, V. J. Differences in parental antecedents of internal–external control in children and in young adulthood. Paper presented at the American Psychological Association Convention, Montreal, 1973.

Crandall, V. J., Katkovsky, W., & Preston, A. Motivational and ability determinants of young children's intellectual academic achievement situations. *Child Development,* 1962, **33,** 643–661.

Crego, C. A. A pattern analytic approach to the measure of modes of expression of psychological differentiation. *Journal of Abnormal Psychology*, 1970, **76**, 194–198.

Cromwell, R. L. Stress, personality, and nursing care in myocardial infarction. Progress choice behavior, and description of parental behavior in schizophrenic and normal subjects. *Journal of Personality*, 1961, **29**, 363–379.

Cromwell, R. L. Stress, personality, and nursing care in myocardial infarction. Progress report, National Institute of Mental Health, 1968.

Crowne, D. P., & Liverant, S. Conformity under varying conditions of personal commitment, *Journal of Abnormal and Social Psychology*, 1963, **66**, 547–555.

Davis, W. L., & Davis, D. E. Internal–external control and attribution of responsibility for success and failure. *Journal of Personality*, 1972, **40**, 123–136.

Davis, W. L., & Phares, E. J. Internal–external control as a determinant of information-seeking in a social influence situation. *Journal of Personality*, 1967, **35**, 547–561.

Davis, W. L., & Phares, E. J. Parental antecedents of internal–external control of reinforcement. *Psychological Reports*, 1969, **24**, 427–436.

Dean, D. G. *Dynamic social psychology: Toward appreciation and replication.* New York: Random House, 1969.

deCharms, R. *Personal causation: The internal affective determinants of behavior.* New York: Academic Press, 1968.

deCharms, R. Personal causation training in the schools. *Journal of Applied Social Psychology*, 1972, **2**, 95–113.

Deever, S. G. Ratings of task oriented expectancy for success as a function of internal control and field independence. *Dissertation Abstracts, Section B*, 1968, **29** (1), 365.

Diamond, M. J., & Shapiro, J. L. Changes in locus of control as a function of encounter group experiences. *Journal of Abnormal Psychology*, 1973, **82**, 514–518.

Doctor, R. Locus of control of reinforcement and responsiveness to social influence. *Journal of Personality*, 1971, **39**, 542–551.

Dua, P. S. Comparison of the effects of behaviorally oriented action and psychotherapy reeducation on intraversion–extraversion, emotionality, and internal–external control. *Journal of Counseling Psychology*, 1970, **17**, 567–572.

DuCette, J., & Wolk, S. Locus of control and extreme behavior. *Journal of Consulting and Clinical Psychology*, 1972, **39**, 253–258.

DuCette, J., & Wolk, S. Cognitive and motivational correlates of generalized expectancies for control. *Journal of Personality and Social Psychology*, 1973, **26**, 420–426.

DuCette, J., Wolk, S., & Soucar, E. Atypical pattern in locus of control and nonadaptive behavior. *Journal of Personality*, 1972, **40**, 287–297.

Dweck, C. S. The role of expectations and attributions on the alleviation of learned helplessness. *Journal of Personality and Social Psychology*, 1975, **31**, 674–685.

Dweck, C. W., & Reppucci, N. D. Learned helplessness and reinforcement responsibility in children. *Journal of Personality and Social Psychology*, 1973, **25**, 109–116.

Efran, J. Some personality determinants of memory for success and failure. Unpublished doctoral dissertation, Ohio State University, 1963.

Elkins, S. Slavery and personality. In B. Kaplan (Ed.), *Studying personality cross-culturally.* New York: Harper & Row, 1961.

Erikson, R. V., & Roberts, A. H. Some ego functions associated with delay of gratification in male delinquents. *Journal of Consulting and Clinical Psychology*, 1971, **36**, 378–382.

Feather, N. T. Some personality correlates of external control. *Australian Journal of Psychology*, 1967, **19**, 253–260.

Feather, N. T. Valence of outcome and expectation of success in relation to task difficulty and perceived locus of control. *Journal of Personality and Social Psychology*, 1968, **7**, 372–386.

Felton, G. S. The experimenter expectancy effect examined as a function of task ambiguity and internal–external control. *Journal of Experimental Research in Personality*, 1971, **5**, 286–294.

Ferster, C. B. A functional analysis of depression. *American Psychologist*, 1973, **28**, 857–870.

Fink, H. C., & Hjelle, L. A. Internal–external control and ideology. *Psychological Reports*, 1973, **33**, 967–974.

Finkelman, J. M., & Glass, D. C. Reappraisal of the relationship between noise and human performance by means of a subsidiary task measure. *Journal of Applied Psychology*, 1970, **54**, 211–213.

Fish, B., & Karabenick, S. A. Relationship between self-esteem and locus of control. *Psychological Reports*, 1971, **29**, 784.

Fitz, R. J. The differential effects of praise and censure on serial learning as dependent on locus of control and field-dependency. *Dissertation Abstracts, International*, 1971, **31**, 4310.

Foulds, M. L. Changes in locus of internal–external control. *Comparative Group Studies*, 1971, **2**, 293–300.

Foulds, M. L., Guinan, J. F., & Warehime, R. G. Marathon group: Changes in perceived locus of control. *Journal of College Student Personnel*, 1974, **15**, 8–11.

Franklin, R. D. Youth's expectancies about internal versus external control of reinforcement related to *N* variables. *Dissertation Abstracts*, 1963, **24**, 1684.

Frazier, E. F. *Black bourgeoisie*. New York: Collier Books, 1962.

Fromm, E. *The art of loving*. New York: Harper & Row, 1956.

Gardner, G. E. Aggression and violence–the enemies of precision learning in children. *American Journal of Psychiatry*, 1971, **128**, 77–82.

Geer, J. H., Davison, G. C., & Gatchel, R. I. Reduction in stress in humans through nonveridical perceived control of aversive stimulation. *Journal of Personality and Social Psychology*, 1970, **16**, 731–738.

Getter, H. A personality determinant of verbal conditioning. *Journal of Personality*, 1966, **34**, 397–405.

Gillis, J. S., & Jessor, R. Effects of brief psychotherapy on belief in internal control: An exploratory study. *Psychotherapy: Theory, Research, and Practice*, 1970, 7, 135–137.

Gilmor, T. M., & Minton, H. L. Internal versus external attribution of task performance as a function of locus of control, initial confidence and success-failure outcome. *Journal of Personality*, 1974, **42**, 159–174.

Glass, D. C., Reim, B., & Singer, J. E. Behavioral consequences of adaptation to controllable and uncontrollable noise. *Journal of Experimental Social Psychology*, 1971, 7, 244–257.

Glass, D. C., & Singer, J. E. *Urban stress*. New York: Academic Press, 1972.

Glass, D. C., Singer, J. E., & Friedman, L. N. Psychic cost of adaptation to an environmental stressor. *Journal of Personality and Social Psychology*, 1969, **12**, 200–210.

Glass, D. C., Singer, J. E., Leonard, H. S., Krantz, D., Cohen, S., & Cummings, H. Perceived control of aversive stimulation and the reduction of stress responses. *Journal of Personality*, 1973, **41**, 577–595.

Golin, S. The effects of stress on the performance of normal and high anxious subjects under chance and skill conditions. *Journal of Abnormal Psychology*, 1974, **83**, 466–472.

Gore, P. S. Individual differences in the prediction of subject compliance to experimenter bias. Unpublished doctoral dissertation, Ohio State University, 1962.

Gore, P. S., & Rotter, J. B. A personality correlate of social action. *Journal of Personality*, 1963, **31**, 58–64.

Gorman, B. S. An observation of altered locus of control following political disappointment. *Psychological Reports*, 1968, **23**, 1094.

Gorman, B. S. A multivariate study of the relationship of cognitive control and cognitive principles to reported daily mood experiences. Unpublished doctoral dissertation, City University of New York, 1971.

Gorsuch, R. L., Henighan, R. P., & Barnard, C. Locus of control: An example of dangers in using children's scales with children. *Child Development,* 1972, **43,** 579–590.

Goss, A., & Morosko, T. E. Relation between a dimension of internal–external control and the MMPI with an alcoholic population. *Journal of Consulting and Clinical Psychology,* 1970, **34,** 189–192.

Gottesfeld, H., & Dozier, G. Changes in feelings of powerlessness in a community action program. *Psychological Reports,* 1966, **19,** 978.

Gozali, J., & Bialer, I. Children's locus of control scale. *American Journal of Mental Deficiency,* 1968, **72,** 622–625.

Gozali, J., Cleary, T. A., Walster, G. W., & Gozali, J. Relationship between the internal–external construct and achievement. *Journal of Educational Psychology,* 1973, **64,** 9–14.

Griffin, J. H. *Black like me.* Boston: Houghton-Mifflin, 1962.

Gurin, G., & Gurin, P. Expectancy theory in the study of poverty. *Journal of Social Issues,* 1970, **26,** 83–104.

Gurin, P., Gurin, G., Lao, R. C., & Beattie, M. Internal–external control in the motivational dynamics of Negro youth. *Journal of Social Issues,* 1969, **25,** 29–53.

Haggard, E. A. Experimental studies in affective processes. *Journal of Experimental Psychology,* 1943, **33,** 257–284.

Hamsher, J. H., Geller, J. D., & Rotter, J. B. Interpersonal trust, internal–external control and the Warren Commission Report. *Journal of Personality and Social Psychology,* 1968, **9,** 210–215.

Harrison, F. I. Relationship between home background, school success, and adolescent attitudes. *Merrill–Palmer Quarterly of Behavior and Development,* 1968, **14,** 331–344.

Harrow, M., & Ferrante, A. Locus of control in psychiatric patients. *Journal of Consulting and Clinical Psychology,* 1969, **33,** 582–589.

Harvey, J. M. Locus of control shift in administrators. *Perceptual and Motor Skills,* 1971, **33,** 980–982.

Hersch, P. D., & Scheibe, K. E. On the reliability and validity of internal–external control as a personality dimension. *Journal of Consulting Psychology,* 1967, **31,** 609–613.

Hiroto, D. S. Validation of the learned helplessness hypothesis with humans. Paper presented at American Psychological Association Convention, Honolulu, 1972.

Hiroto, D. S. Locus of control and learned helplessness. *Journal of Experimental Psychology,* 1974, **102,** 187–193.

Hiroto, D. S., & Seligman, M. E. P. Generality of learned helplessness in man. *Journal of Personality and Social Psychology,* 1975, **31,** 311–327.

Hjelle, L. A. Social desirability as a variable in the locus of control scale. *Psychological Reports,* 1971, **28,** 807–816.

Hokanson, J. E., DeGood, D. E. Forrest, M. S., & Brittain, T. M. Availability of avoidance behaviors in modulating vascular stress responses. *Journal of Personality and Social Psychology,* 1971, **19,** 60–68.

Houston, B. K. Control over stress, locus of control, and response to stress. *Journal of Personality and Social Psychology,* 1972, **21,** 249–255.

Hsieh, T. T., Shybut, J., & Lotsof, E. J. Internal versus external control and ethnic group membership. *Journal of Consulting and Clinical Psychology,* 1969, **33,** 122–124.

James, W. H. Internal versus external control of reinforcement as a basic variable in learning theory. Unpublished doctoral dissertation, Ohio State University, 1957.

James, W. H. The application of social learning theory to educational processes. Paper presented at a meeting of the Society for Research in Child Development, Minneapolis, 1965.

James, W. H., & Rotter, J. B. Partial and 100 percent reinforcement under chance and skill conditions. *Journal of Experimental Psychology*, 1958, **55**, 397–403.

James, W. H., Woodruff, A. B., & Werner, W. Effect of internal and external control upon changes in smoking behavior. *Journal of Consulting Psychology*, 1965, **29**, 127–129.

Jessor, R., Graves, T. D., Hanson, R. C., & Jessor, S. *Society, personality, and deviant behavior.* New York: Holt, Rinehart, & Winston, 1968.

Joe, V. C. A review of the internal–external control construct as a personality variable. *Psychological Reports*, 1971, **28**, 619–640.

Joe, V. C. Social desirability and the I–E scale. *Psychological Reports*, 1972, **30**, 44–46.

Joe, V. C., & Jahn, J. C. Factor structure of the Rotter I–E scale. *Journal of Clinical Psychology*, 1973, **29**, 66–68.

Johnson, C. D., & Gormly, J. Academic cheating: The contribution of sex, personality, and situational variables. *Developmental Psychology*, 1972, **6**, 320–325.

Johnson, R. C., Ackerman, J. M., Frank, H., & Fionda, A. J. Resistance to temptation and guilt following yielding and psychotherapy. *Journal of Consulting and Clinical Psychology*, 1968, **32**, 169–175.

Julian, J. W., & Katz, S. B. Internal versus external control and the value of reinforcement. *Journal of Personality and Social Psychology*, 1968, **76**, 43–48.

Karabenick, S. A. Valence of success and failure as a function of achievement motives and locus of control. *Journal of Personality and Social Psychology*, 1972, **21**, 101–110.

Katkovsky, W., Crandall, V. C., & Good, S. Parental antecedents of children's beliefs in internal–external control of reinforcement in intellectual achievement situations. *Child Development*, 1967, **28**, 765–776.

Katz, I. The socialization of academic motivation in minority group children. In D. Levine (Ed.) *Nebraska symposium on motivation.* Lincoln, Nebraska: University of Nebraska Press, 1967, pp. 133–191.

Kelman, H. C., & Lawrence, L. H. Assignment of responsibility in the case of Lt. Calley: Preliminary report on a national survey. *Journal of Social Issues*, 1972, 28 (1), 177–212.

Kiehlbauch, J. B. Selected changes over time in internal–external control expectancies in a reformatory population. Unpublished doctoral dissertation, Kansas State University, 1968.

Kilpatrick, D. G., Dubin, W. R., & Marcotte, D. B. Personality, stress of the medical education process, and changes in affective mood state. *Psychological Reports*, 1974, **34**, 1215–1223.

Kirscht, J. P. Perception of control and health beliefs. *Canadian Journal of Behavioral Science*, 1972, **4**, 225–237.

Kish, G. B., Solberg, K. B., & Uecker, A. E. Locus of control as a factor influencing patients-perceptions of ward atmosphere. *Journal of Clinical Psychology*, 1971, **27**, 287–289.

Kleiber, D., Veldman, D. J., & Menaker, S. L. The multidimensionality of locus of control. Paper presented at Eastern Psychological Association Convention, Washington, D.C., 1973.

Kobler, A. L., & Stotland, E. *The end of hope.* New York: Free Press, 1964.

Koestler, A. *The ghost in the machine.* New York: Macmillan, 1967.

Lao, R. C. Internal–external control and competent and innovative behavior among Negro college students. *Journal of Personality and Social Psychology*, 1970, **14**, 263–270.

Lefcourt, H. M. Risk taking in Negro and white adults. *Journal of Personality and Social Psychology*, 1965, **2**, 765–770.

Lefcourt, H. M. Internal–external control of reinforcement: A review. *Psychological Bulletin*, 1966, **65**, 206–220. (a)

Lefcourt, H. M. Belief in personal control: A goal for psychotherapy. *Journal of Individual Psychology*, 1966, **22**, 185–195. (b)

Lefcourt, H. M. The effects of cue explication upon persons maintaining external control expectancies. *Journal of Personality and Social Psychology,* 1967, **5**, 372–378.

Lefcourt, H. M. Recent developments in the study of locus of control. In B. A. Maher (Ed.), *Progress in experimental research in personality.* Vol. 6. New York: Academic Press, 1972.

Lefcourt, H. M. The function of the illusions of control and freedom. *American Psychologist,* 1973, **28**, 417–425.

Lefcourt, H. M. Locus of control and the acceptance of one's reinforcement experience. Paper presented at the World Congress of Sociology, Toronto, 1974.

Lefcourt, H. M., Antrobus, P. M., & Hogg, E. Humor response and humor production as a function of locus of control, field dependence, and type of reinforcements. *Journal of Personality,* 1974, **42**, 632–651.

Lefcourt, H. M., Gronnerud, P., & McDonald, P. Cognitive activity and hypothesis formation during a double entendre word association test as a function of locus of control and field dependence. *Canadian Journal of Behavioral Science,* 1973, **5**, 161–173.

Lefcourt, H. M., Hogg, E., & Sordoni, Carol. Locus of control, field dependence, and conditions arousing objective versus subjective self-awareness. *Journal of Research in Personality,* 1975, **9**, 21–36.

Lefcourt, H. M., Hogg, E., Struthers, S., & Holmes, C. Causal attributions as a function of locus of control, initial confidence, and performance outcomes. *Journal of Personality & Social Psychology,* 1975, **32**, 391–397.

Lefcourt, H. M., & Ladwig, G. W. The American Negro: A problem in expectancies. *Journal of Personality and Social Psychology,* 1965, **1**, 377–380.   (a)

Lefcourt, H. M., & Ladwig, G. W. The effect of reference group upon Negroes' task persistence in a biracial competitive game. *Journal of Personality and Social Psychology,* 1965, **1**, 668–671.   (b)

Lefcourt, H. M., & Ladwig, G. W. Alienation in Negro and white reformatory inmates. *Journal of Social Psychology,* 1966, **68**, 152–157.

Lefcourt, H. M., Lewis, L., & Silverman, I. W. Internal versus external control of reinforcement and attention in decision-making tasks. *Journal of Personality,* 1968, **36**, 663–682.

Lefcourt, H. M., & Siegel, J. Reaction time behavior as a function of internal–external control of reinforcement and control of test administration. *Canadian Journal of Behavioral Science,* 1970, **2**, 253–266.   (a)

Lefcourt, H. M., & Siegel, J. Reaction time performance as a function of field dependence and autonomy in test administration. *Journal of Abnormal Psychology,* 1970, **76**, 475–481.   (b)

Lefcourt, H. M., Sordoni, Carl, & Sordoni, Carol. Locus of control, field dependence, and the expression of humor. *Journal of Personality,* 1974, **42**, 130–143.

Lefcourt, H. M., & Steffy, R. A. Level of aspiration, risk-taking behavior, and projective test performance: A search for coherence. *Journal of Consulting and Clinical Psychology,* 1970, **34**, 193–198.

Lefcourt, H. M., & Telegdi, M. Perceived locus of control and field dependence as predictors of cognitive activity. *Journal of Consulting and Clinical Psychology,* 1971, **37**, 53–56.

Lefcourt, H. M., & Wine, J. Internal versus external control of reinforcement and the deployment of attention in experimental situations. *Canadian Journal of Behavioral Science,* 1969, **1**, 167–181.

Lessing, E. E. Racial differences in indices of ego functioning relevant to academic achievement. *Journal of Genetic Psychology,* 1969, **115**, 153–167.

Levenson, H. Distinctions within the concept of internal–external control. Paper presented at the American Psychological Association Convention, Washington, D.C., 1972.

Levenson, H. Multidimensional locus of control in psychiatric patients. *Journal of Consulting and Clinical Psychology,* 1973, **41**, 397–404.   (a)

Levenson, H. Perceived parental antecedents of internal, powerful others and chance locus of control orientations. *Developmental Psychology,* 1973, 9, 268–274. (b)

Levenson, H. Activism and powerful others: Distinctions within the concept of internal–external control. *Journal of Personality Assessment,* 1974, 38, 377–383.

Levy, L. H. Expectancy for locus of control of reinforcement and perception of orderliness. *Perceptual and Motor Skills,* 1967, 25, 781–786.

Lewin, K. Psycho-sociological problems of a minority group. *Character and Personality,* 1940, 8, 176–187.

Lewinsohn, P. Clinical and theoretical aspects of depression. Paper presented at the Georgia Symposium in Experimental Clinical Psychology, 1972.

Lewis, O. *Children of Sanchez.* New York: Random House, 1961.

Lichtenstein, E., & Keutzer, C. S. Further normative and correlational data on the internal–external (I–E) control of reinforcement scale. *Psychological Reports,* 1967, 21, 1014–1016.

Lipp, L., Kolstoe, R., James, W., & Randall, H. Denial of disability and internal control of reinforcement: A study using a perceptual defense paradigm. *Journal of Consulting and Clinical Psychology,* 1968, 32, 72–75.

Lotsof, E. J., James, W. H., Drucker, R., & Blount, W. Perceptual discrimination, task structure and locus of control. Paper presented at Midwestern Psychological Association, St. Louis, 1964.

Lottman, T. J., & DeWolfe, A. S. Internal versus external control in reactive and process schizophrenia. *Journal of Consulting and Clinical Psychology,* 1972, 39, 344.

Ludwigsen, K., & Rollins, H. Recognition of random forms as a function of source of cue, perceived locus of control, and socioeconomic level. Paper presented at Southeastern Psychological Association Convention, Miami Beach, 1971.

MacDonald, A. P., & Hall, J. Internal–external locus of control and perception of disability. *Journal of Consulting and Clinical Psychology,* 1971, 36, 338–343.

MacDonald, A. P., & Tseng, M. S. Dimensions of internal versus external control revisited. Unpublished manuscript, University of West Virginia, 1971.

Maher, B. *Principles of psychopathology.* New York: McGraw-Hill, 1966.

Mahrer, A. R. The role of expectancy in delayed reinforcement. *Journal of Experimental Psychology,* 1956, 52, 101–105.

Martin, R. D., & Shepel, L. F. Locus of control and discrimination ability with lay counselors. *Journal of Consulting and Clinical Psychology,* 1974, 42, 741.

Maslow, A. H. *Motivation and personality.* New York: Harper & Row, 1954.

Masters, J. C. Treatment of adolescent rebellion by the reconstrual of stimuli. *Journal of Consulting Psychology,* 1970, 35, 213–216.

May, R. *Love and will.* New York: Norton, 1969.

McArthur, L. A. Luck is alive and well in New Haven. *Journal of Personality and Social Psychology,* 1970, 16, 316–318.

McClelland, D. C., Atkinson, J. W., Clark, R. A., & Lowell, E. L. *The achievement motive.* New York: Appleton-Century-Crofts, 1953.

McGhee, P. E., & Crandall, V. C. Beliefs in internal–external control of reinforcement and academic performance. *Child Development,* 1968, 39, 91–102.

McNair, D. M., Lorr, M., & Droppleman, L. *The profile of mood states.* Educational and Industrial Testing Service, San Diego, 1971.

Melges, F. T., & Weisz, A. E. The personal future and suicidal ideation. *Journal of Nervous and Mental Disease,* 1971, 153, 244–250.

Menninger, K. *The vital balance.* New York: Viking, 1963.

Messer, S. B. The relation of internal–external control to academic performance. *Child Development,* 1972, 43, 1456–1462.

Meyer, W. U. Selbstverantwortlichkeit und Leistungs motivation. Unpublished doctoral dissertation, Ruhr University. Bochum, Germany, 1970.

Midlarski, E. Aiding under stress: The effects of competence, dependency, visibility, and fatalism. *Journal of Personality*, 1971, **39**, 132–149.

Milgram, S. Behavioral study of obedience. *Journal of Abnormal and Social Psychology*, 1963, **67**, 371–378.

Milgram, S. Some conditions of obedience and disobedience to authority. In I. D. Steiner & M. Fishbein (Eds.), *Current studies in social psychology.* New York: Holt, Rinehart, & Winston, 1965.

Miller, A. G., & Minton, H. L. Machiavelliansm, internal–external control and the violation of experimental instructions. *Psychological Record,* 1969, **19**, 369–380.

Miller, W. R., & Seligman, M. E. P. Depression and the perception of reinforcement. *Journal of Abnormal Psychology,* 1973, **82**, 62–73.

Mirels, H. L. Dimensions of internal versus external control. *Journal of Consulting and Clinical Psychology,* 1970, **34**, 226–228.

Mirels, H. L., & Garrett, J. B. The Protestant ethic as a personality variable. *Journal of Consulting and Clinical Psychology,* 1971, **36**, 40–44.

Mischel, W. Preference for delayed reinforcement and social responsibility. *Journal of Abnormal and Social Psychology,* 1961, **62**, 1–7.

Mischel, W. Theory and research on the antecedents of self-imposed delay of reward. In B. A. Maher (Ed.), *Progress in experimental personality research.* Vol. 3. New York: Academic Press, 1966.

Mischel, W. Towards a cognitive social learning reconceptualization of personality. *Psychological Review,* 1973, **80**, 252–283.

Mischel, W., Zeiss, R., & Zeiss, A. Internal–external control and persistence: Validation and implications of the Stanford preschool internal–external scale. *Journal of Personality and Social Psychology,* 1974, **29**, 265–278.

Mowrer, O. H. *Learning theory and personality dynamics.* New York: Ronald Press, 1950.

Mowrer, O. H., & Viek, P. An experimental analogue of fear from a sense of helplessness. *Journal of Abnormal and Social Psychology,* 1948, **43**, 193–200.

Myrdal, G. *An American dilemma.* New York: Harper & Row, 1944.

Naditch, M. Putting the value back into expectancy value theory. Paper presented at Eastern Psychological Association Convention, Washington, D.C., 1973.

Nowicki, S. Predicting academic achievement of females from locus of control orientation: Some problems and some solutions. Paper presented at American Psychological Association Convention, Montreal, 1973.

Nowicki, S., & Barnes, J. Effects of a structured camp experience on locus of control orientation. *Journal of Genetic Psychology,* 1973, *122,* 247–252.

Nowicki, S., & Roundtree, J. Correlates of locus of control in secondary school age students. Unpublished manuscript, Emory University, 1971.

Nowicki, S., & Strickland, B. R. A locus of control scale for children. *Journal of Consulting and Clinical Psychology,* 1973, **40**, 148–154.

Odell, M. Personality correlates of independence and conformity. Unpublished master's thesis. Ohio State University, 1959.

Overmier, J. B., & Seligman, M. E. P. Effects of inescapable shock upon subsequent escape and avoidance responding. *Journal of Comparative and Physiological Psychology,* 1967, **63**, 23–33.

Owens, M. W. Disability–minority and social learning. Unpublished master's thesis, West Virginia University, 1969.

Palmer, R. D. Parental perception and perceived locus of control in psychopathology. *Journal of Personality,* 1971, **3**, 420–431.

Parsons, O. A., & Schneider, J. M. Locus of control in university students from eastern and western societies. *Journal of Consulting and Clinical Psychology,* 1974, **42,** 456–461.

Penk, W. Age changes and correlates of internal–external locus of control scales. *Psychological Reports,* 1969, **25,** 856.

Pervin, L. A. The need to predict and control under conditions of threat. *Journal of Personality,* 1963, **31,** 570–585.

Pettigrew, T. F. Social evaluation theory: Convergences and applications. In D. Levine (Ed.), *Nebraska symposium on motivation.* Lincoln, Nebraska: University of Nebraska Press, 1967.

Phares, E. J. Expectancy changes in skill and chance situations. *Journal of Abnormal and Social Psychology,* 1957, **54,** 339–342.

Phares, E. J. Differential utilization of information as a function of internal–external control. *Journal of Personality,* 1968, **36,** 649–662.

Phares, E. J. Internal–external control and the reduction of reinforcement value after failure. *Journal of Consulting and Clinical Psychology,* 1971, **37,** 386–390.

Phares, E. J. *Locus of control: A personality determinant of behavior.* Morristown, New Jersey: General Learning Press, 1973.

Phares, E. J., Ritchie, D. E., & Davis, W. L. Internal–external control and reaction to threat. *Journal of Personality and Social Psychology,* 1968, **10,** 402–405.

Phares, E. J., Wilson, K. G., & Klyver, N. W. Internal–external control and the attribution of blame under neutral and distractive conditions. *Journal of Personality and Social Psychology,* 1971, **18,** 285–288.

Pines, H. A. An atributional analysis of locus of control orientation and source of informational dependence. Unpublished manuscript, State University of New York at Buffalo, 1974.

Pines, H. A., & Julian, J. W. Effects of task and social demands on locus of control differences in information processing. *Journal of Personality,* 1972, **40,** 407–416.

Platt, E. S. Internal–external control and changes in expected utility as predictors of the change in cigarette smoking following role-playing. Paper presented at the Eastern Psychological Association Convention, Philadelphia, 1969.

Platt, J. J., & Eisenman, R. Internal–external control of reinforcement, time perspective, adjustment, and anxiety. *Journal of General Psychology,* 1968, **79,** 121–128.

Powell, A., & Vega, M. Correlates of adult locus of control. *Psychological Reports,* 1972, **30,** 455–460.

Ray, W. J., & Katahn, M. Relation of anxiety to locus of control. *Psychological Reports,* 1968, **23,** 1196.

Reid, D., & Ware, E. E. Multidimensionality of internal versus external control: Addition of a third dimension and non-distinction of self versus others. *Canadian Journal of Behavioral Science,* 1974, **6,** 131–142.

Reid, D., & Ware, E. E. Multidimensionality of internal–external control: Implications for past and future research. *Canadian Journal of Behavioral Science,* 1973, **5,** 264–271.

Reim, B., Glass, D. C., & Singer, J. E. Behavioral consequences of exposure to uncontrollable and unpredictable noise. *Journal of Applied Social Psychology,* 1971, **1,** 44–56.

Reimanis, G. Effects of experimental IE modification techniques and home environment variables on IE. Paper presented at the American Psychological Association Convention, Washington, D.C., 1971.

Richter, C. P. The phenomenon of unexplained sudden death in animals and man. In H. Feifel (Ed.), *The meaning of death.* New York: McGraw-Hill, 1959.

Riessman, F. *The culturally deprived child.* New York: Harper & Row, 1962.

Ritchie, D. E., & Phares, E. J. Attitude change as a function of internal–external control and communicator status. *Journal of Personality,* 1969, *37,* 429–443.

Rose, A. *The negro in America.* New York: Harper & Row, 1948.

Rosenzweig, S. The picture association method and its application in a study of reactions to frustration. *Journal of Personality,* 1945, **14,** 3–23.

Rotter, J. B. *Social learning and clinical psychology.* Englewood Cliffs, New Jersey: Prentice-Hall, 1954.

Rotter, J. B. The role of the psychological situation in determining the direction of human behavior. In M. R. Jones (Ed.), *Nebraska symposium on motivation,* Lincoln, Nebraska: University of Nebraska Press, 1955.

Rotter, J. B. Some implications of a social learning theory for the prediction of goal directed behavior from testing procedures. *Psychological Review,* 1960, **67,** 301–316.

Rotter, J. B. *Clinical psychology.* Englewood Cliffs, New Jersey: Prentice-Hall, 1964.

Rotter, J. B. Generalized expectancies for internal versus external control of reinforcement. *Psychological Monographs,* 1966, **80** (Whole No. 609).

Rotter, J. B. External control and internal control. *Psychology Today,* 1971, **5,** 37–59.

Rotter, J. B. Some problems and misconceptions related to the construct of internal vs. external control of reinforcement. *Journal of Consulting and Clinical Psychology,* 1975, **48,** 56–67.

Rotter, J. B., Chance, J. E., & Phares, E. J. *Applications of a social learning theory of personality.* New York: Holt, Rinehart, & Winston, 1972.

Rotter, J. B., Liverant, S., & Crowne, D. P. The growth and extinction of expectancies in chance controlled and skilled tasks. *Journal of Psychology,* 1961, **52,** 161–177.

Rotter, J. B., & Mulry, R. C. Internal versus external control of reinforcements and decision time. *Journal of Personality and Social Psychology,* 1965, **2,** 598–604.

Rotter, J. B., Seeman, M., & Liverant, S. Internal versus external control of reinforcement: A major variable in behavior theory. In N. F. Washburne (Ed.), *Decisions, values and groups.* Vol. 2. Oxford: Pergamon Press, 1962, 473–516.

Rotter, J. B., Simmons, W. L., & Holden, K. B. Some correlates of a general attitude towards the locus of control of reinforcement. Unpublished manuscript, The Ohio State University, 1962.

Ryckman, R. M., & Sherman, M. F. Relationship between self-esteem and internal–external control for men and women. *Psychological Reports,* 1973, *32,* 1106.

Schaefer, E. S., & Bell, R. Q. Development of a parental attitude research instrument. *Child Development,* 1958, **29,** 339–361.

Schneider, J. M., & Parsons, O. A. Categories of the locus of control scale and cross-cultural comparisons in Denmark and the United States. *Journal of Cross-Cutural Psychology,* 1970, **1,** 131–138.

Scott, D. The Negro and the enlisted man: An analogy. *Harper's Magazine,* 1962, **225,** 16–21.

Seeman, M. Alienation and social learning in a reformatory. *American Journal of Sociology,* 1963, **69,** 270–284.

Seeman, M., & Evans, J. W. Alienation and learning in a hospital setting. *American Sociological Review,* 1962, **27,** 772–783.

Seligman, M. E. P. Chronic fear produced by unpredictable shock. *Journal of Comparative and Physiological Psychology,* 1968, **66,** 402–411.

Seligman, M. E. P., & Maier, S. F. Failure to escape traumatic shock. *Journal of Experimental Psychology,* 1967, **74,** 1–9.

Seligman, M. E. P., Maier, S. F., & Geer, J. H. The alleviation of learned helplessness in the dog. *Journal of Abnormal and Social Psychology,* 1968, **73,** 256–262.

Seligman, M. E. P., Maier, S. F., & Solomon, R. L. Unpredictable and uncontrollable aversive events. In F. R. Brush (Ed.), *Aversive conditioning and learning.* New York: Academic Press, 1969.

Shaw, R. L., & Uhl, N. P. Relationship between locus of control scores and reading achievement of black and white second grade children from two socioeconomic levels. Paper presented at the Southeastern Psychological Association Convention, New Orleans, 1969.

Sherman, S. Internal–external control and its relationship to attitude change under different social influence techniques. *Journal of Personality and Social Psychology,* 1973, **23,** 23–29.

Shipe, D. Impulsivity and locus of control as predictors of achievement and adjustment in mildly retarded and borderline youth. *American Journal of Mental Deficiency,* 1971, **76,** 12–22.

Shore, R. E. Parental determinants of boys internal–external control. Unpublished doctoral dissertation, Syracuse University, 1967.

Shybut, J. Time perspective, internal versus external control and severity of psychological disturbance. *Journal of Clinical Psychology,* 1968, **24,** 312–315.

Sims, J. H., & Baumann, D. D. The tornado threat: Coping styles of the north and south. *Science,* 1972, **176,** 1386–1392.

Singer, E. *Key concepts in psychotherapy.* New York: Random House, 1965.

Skinner, B. F. *Beyond freedom and dignity.* New York: Knopf, 1971.

Smith, C. E., Pryer, M. W., & Distefano, M. K. Internal–external control and severity of emotional impairment among psychiatric patients. *Journal of Clinical Psychology,* 1971, **27,** 449–450.

Smith, R. E. Changes in locus of control as a function of life crisis resolution. *Journal of Abnormal Psychology,* 1970, **75,** 328–332.

Snyder, C. R., & Larson, G. R. A further look at student acceptances of general personality interpretations. *Journal of Consulting and Clinical Psychology,* 1972, **38,** 384–388.

Solomon, D., Houlihan, K. A., & Parelius, R. Intellectual achievement responsibility in negro and white children. *Psychological Reports,* 1969, **24,** 479–483.

Sordoni, Carl. Personality and situational determinants of humor appreciation and humor production. Unpublished doctoral dissertation, University of Waterloo, 1975.

Staub, E., Tursky, B., & Schwartz, G. E. Self-control and predictability: Their effects on reactions to aversive stimulation. *Journal of Personality and Social Psychology,* 1971, **18,** 157–162.

Steiner, I. D. Perceived freedom. In L. Berkowitz (Ed.), *Advances in experimental social psychology.* Vol. 5. New York: Academic Press, 1970, pp. 187–248.

Stephens, M. W. Parent behavior antecedents, cognitive correlates and multidimensionality of locus of control in young children. Paper presented at the American Psychological Association Convention, Montreal, 1973.

Stephens, M. W., & Delys, P. A locus of control measure for preschool children. *Developmental Psychology,* 1973, **9,** 55–65.   (a)

Stephens, M. W., & Delys, P. External control expectancies among disadvantaged children at preschool age. *Child Development,* 1973, **44,** 670–674.   (b)

Strickland, B. R. Individual differences in verbal conditioning, extinction, and awareness. *Journal of Personality,* 1970, **38,** 364–378.

Strickland, B. R. Delay of gratification as a function of race of the experimenter. *Journal of Personality and Social Psychology,* 1972, **22,** 108–112.

Strickland, B. R. The prediction of social action from a dimension of internal–external control. *Journal of Social Psychology,* 1965, **66,** 353–358.

Strodtbeck, F. L. Family interaction, values and achievement. In D. C. McClelland (Ed.), *Talent and society.* New York: Van Nostrand, 1958.

Sullivan, H. S. *The psychiatric interview.* New York: Norton, 1954.

Thomas, L. E. The I–E scale, ideological bias and political participation. *Journal of Personality,* 1970, **38,** 273–286.

Time Magazine, Sept 20, 1971. In *Annual Editions Readings in Psychology 1972* Guilford, Connecticut: Dushkin Publishing Group, 1972. Pp. 6–13.

Tolor, A., & Reznikoff, M. Relation between insight, repression–sensitization, internal–external control, and death anxiety. *Journal of Abnormal Psychology,* 1967, **72,** 426–430.

Valins, S., & Nisbett, R. E. *Some implications of the attribution processes for the development and treatment of emotional disorders.* New York: General Learning Press, 1971.

Viney, L. L. Multidimensionality of perceived locus of control: Two replications. *Journal of Consulting and Clinical Psychology,* 1974, **42,** 463–464.

Walls, R. T., & Miller, J. J. Delay of gratification in welfare and rehabilitation clients. *Journal of Counseling Psychology,* 1970, **4,** 383–384.

Walls, R. T., & Smith, T. S. Development of preference for delayed reinforcement in disadvantaged children. *Journal of Educational Psychology,* 1970, **61,** 118–123.

Warehime, R. G., & Woodson, M. Locus of control and immediate affect states. *Journal of Clinical Psychology,* 1971, **27,** 443–444.

Watson, D. Relationship between locus of control and anxiety. *Journal of Personality and Social Psychology,* 1967, **6,** 91–92.

Watson, D., & Baumal, E. Effects of locus of control and expectation of future control upon present performance. *Journal of Personality and Social Psychology,* 1967, **6,** 212–215.

Watson, J. S. The development and generalization of "contingency awareness" in early infancy: Some hypotheses. *Merrill–Palmer Quarterly of Behavior and Development,* 1966, **12,** 123–135.

Watson, J. S. Memory and "contingency analysis" in infant learning. *Merrill–Palmer Quarterly of Behavior and Development,* 1967, **13,** 55–76.

Watson, J. S. Smiling, cooing and "the game." *Merrill–Palmer Quarterly of Behavior and Development,* 1972, **18,** 323–339.

Watson, J. S., & Ramey, C. T. Reactions to response contingent stimulation in early infancy. *Merrill–Palmer Quarterly of Behavior and Development,* 1972, **18,** 219–227.

Weiner, B. New conceptions in the study of achievement motivation. In B. A. Maher (Ed.), *Progress in experimental personality research.* Vol. 5. New York: Academic Press, 1970, pp. 67–109.

Weiner, B. *Theories of motivation: From mechanism to cognition.* Chicago: Markham, 1972.

Weiner, B., Friese, I., Kukla, A., Reed, L., Rest, S., & Rosenbaum, R. M. *Perceiving the causes of success and failure.* New York: General Learning Press, 1971.

Weiner, B. Achievement motivation as conceptualized by an attribution theorist. In B. Weiner (Ed.), *Attribution theory, achievement motivation.* New York: General Learning Press, 1974.

Weiner, B., Heckhausen, H., Meyer, W. U., & Cook, R. E. Causal ascriptions and achievement motivation. *Journal of Personality and Social Psychology,* 1972, **21,** 239–248.

Weiss, J. M. Effects of coping behavior in different warning signal conditions on stress pathology in rats. *Journal of Comparative and Physiological Psychology,* 1971, **77,** 1–13.

White, R. W. The experience of efficacy in schizophrenia. *Psychiatry: Journal for the Study of Interpersonal Processes,* 1965, **28,** 199–211.

Williams, J. G., & Stack, J. J. Internal–external control as a situational variable in determining information-seeking by negro students. *Journal of Consulting and Clinical Psychology,* 1972, **39,** 187–193.

Witkin, H., Dyk, R. B., Faterson, H. F., Goodenough, D. R., & Karp, S. A. *Psychological differentiation.* New York: Wiley, 1962.

Wolfgang, A., & Potvin, R. Internality as a determinant of classroom participation and academic performance among elementary students. Paper presented at the American Psychological Association Convention, Montreal, 1973.

Wolk, S., & DuCette, J. The moderating effect of locus of control in relation to achievement-motivation variables. *Journal of Personality,* 1973, **41,** 59–70.

Wolk, S., & DuCette, J. Intentional performance and incidental learning as a function of personality and task directions. *Journal of Personality and Social Psychology,* 1974, **29,** 90–101.

Wortman, C. B. Some determinants of perceived control. *Journal of Personality and Social Psychology,* 1975, **31,** 282–294.

Zytkoskee, A., Strickland, B. R., & Watson, J. Delay of gratification and internal versus external control among adolescents of low socioeconomic status. *Developmental Psychology,* 1971, **4,** 93–98.

Zytowski, D. G. Internal–external control of reinforcement and the Strong Vocational Interest Blank. *Journal of Consulting Psychology,* 1967, **14,** 177–179.

Author Index

The numbers in *italics* refer to the pages on which the complete references are listed.

AUTHOR INDEX **207**

Subject Index

S

Schizophrenia, 90–91
Self reliance, 60
Skill, 31–34, 47, 54–58, 143
Slavery, 20
Smoking, 43–44
Social class, 17, 22–24
Social cue effects, 144–145
Social desirability, 151
Social influence, 53
Success, 31–35
Sudden death, 8–11
Suggestibility, 144
Suicide, 11, 76–77, 92

T

Task relevant thoughts, 58
Thematic Apperception Test (TAT), 42, 59
Time judgments, 76–80

U

Ulcers, 13–14
Unusual shifts, *see* Level of aspiration

V

Value expectancy, *see* Interactive models
Verbal conditioning, 42–46
Verbal fluency, 145
Vitality, 152–153

W

Will, 2–3, 8–10
Word association, 61–62

Y

Yoked controls, 7, 13–14